软件工程系列教材

软件项目实践案例教程

毛玉萃 邱少明 杨文艳 秦静 编著

清华大学出版社
北京

内容简介

本书以几个典型软件项目案例的开发过程为主线，首先介绍案例开发中应用软件项目开发的方法（原型法、结构化系统方法和面向对象方法）、采用的软件架构（三层架构——表现层、业务逻辑层和数据访问层，MVC——模型-视图-控制器）、软件开发平台和开发工具（SQL Server 2008、MySQL、.NET、Java及Java环境）；然后介绍五个典型的教学案例（从项目的规划、系统分析、系统设计，直至系统实现）；最后针对一个实际项目的完整开发过程进行详细介绍。每个案例和实际项目都提供了完整的可运行系统。

本书共有8章，可作为软件项目开发人员的参考书，也可作为普通高等院校计算机科学与技术、软件工程等相关专业的教学和实践用书或参考书，还可作为培训机构的项目实践培训用书以及软件项目开发爱好者的参考书。

本书封面贴有清华大学出版社防伪标签，无标签者不得销售。
版权所有，侵权必究。举报：010-62782989，beiqinquan@tup.tsinghua.edu.cn。

图书在版编目（CIP）数据

软件项目实践案例教程/毛玉萃等编著. --北京：清华大学出版社，2014（2023.8重印）
软件工程系列教材
ISBN 978-7-302-36350-7

Ⅰ.①软… Ⅱ.①毛… Ⅲ.①软件开发—项目管理—教材 Ⅳ.①TP311.52

中国版本图书馆CIP数据核字（2014）第099143号

责任编辑：白立军　徐跃进
封面设计：傅瑞学
责任校对：焦丽丽
责任印制：宋　林

出版发行：清华大学出版社
网　　址：http://www.tup.com.cn, http://www.wqbook.com
地　　址：北京清华大学学研大厦A座　　　　邮　编：100084
社 总 机：010-83470000　　　　　　　　　邮　购：010-62786544
投稿与读者服务：010-62776969, c-service@tup.tsinghua.edu.cn
质量反馈：010-62772015, zhiliang@tup.tsinghua.edu.cn
课件下载：http://www.tup.com.cn, 010-83470236

印 装 者：三河市人民印务有限公司
经　　销：全国新华书店
开　　本：185mm×260mm　　　印　张：24.25　　　字　数：561千字
版　　次：2014年9月第1版　　　　　　　　　印　次：2023年8月第7次印刷
定　　价：69.00元

产品编号：055631-03

前言

软件工程系列教材

目前大学生就业形势不容乐观,尤其是计算机相关专业的学生。虽然相应岗位目前需求量很大,每年相关专业的毕业生也很多,但很多公司和企业依旧招不到合适的人才,许多相关专业的大学毕业生找不到合适的工作,产生这样结果的原因主要是现在的应届毕业生自身的技能无法满足企业的任职需求,现在的企业不可能花大量的时间和金钱去培养新员工。

学校为此对相关专业的培养方案进行了调整,增设或加强了软件项目开发实践环节的学时,让学生通过专业的系统的实战,提高自己的实践动手操作能力,从而提高和加强学生软件项目开发实践方面的能力,达到学以致用、积累经验、增长见识的目的,提高学生的专业素质和就业能力,以便顺利进入工作岗位。

在软件项目实践的教学过程中,选择一种合适的教材比较难,教材基本都是教师自己编写的讲义,经过几轮的教学之后,讲义几易其稿不断完善,其中的案例不断修订完善、筛选,优中选优,有的案例还是作者开发的实际应用系统。作者在讲义的基础上,再次进行修改和完善,形成了这本书。

本书内容的安排与组织情况如下:第1章为系统开发方法简介,介绍开发案例和实际系统中用到的软件项目开发方法,主要是较成熟和常用的原型法和结构化开发方法,以及目前比较流行的面向对象的开发方法。第2章为开发技术和开发环境简介,对在开发案例和实际系统中用到的数据库(SQL Server 2008和MySQL)和开发平台(.NET、Java语言和开发环境MyEclipse 8.6)进行了简单介绍,对采用的系统架构(三层架构和MVC架构)进行了简单介绍。第3章运用Java类和面向对象基本思想完成一个简单的ATM存取款管理系统的设计与实现,目的在于使学生掌握MyEclipse开发环境的安装、使用以及用类和对象实现系统逻辑。第4章运用Java Swing和事件处理机制以及JDBC编程技术实现一个比较实用的网络考试系统项目,包括学生考试、教师出题等功能,目的在于使学生掌握 MySQL 数据库的安装、使用及在 Java 程序中实现数据访问和维护。第 5 章是网上灯饰店的研究与实现,采用的开发环境是 Visual Studio 2010,数据库为 SQL Server 2008,对网上灯饰店的规划(需求分析、目标设定、可行性分析)、分析(业务流程分析、数据流程分析、数据分析)、设计(功能结构设计、数据库设计、功能详细设计)和实现(系统运行环境、数据库建立与连接、总体框架、数据访问层、业务逻辑层、表示层、公共库、接口等)进行详细介绍和描述。第 6 章是家具网站的研究与实现,采用的开发环境是 MyEclipse 8.6,数据库为 MySQL,对网上灯饰店的规划(需求分析、可行性分析)、分析(业务流程分析、数据流程分析、数据分析)、设计(功能结构设计、数据库设计、功能详细设计)和实现(系统运行环境、数据库建立与连接、总体框架、数据访问层、业务逻辑层、表示层)进行了详细介绍和描述。第 7 章以一个简单通用的办公自动化模拟系统为例,使用面向对象的分析和设计方法,建立了系统 UML 模型;采用 MVC 三层架构,完成了系统实现;并且以员工个人信息管理模块为例,具体说明了代码的开发过程。第 8 章详细介绍一个实际系统——大学毕业(论文)设计管理网站的研究和实现。

  本书由毛玉萃、邱少明、杨文艳和秦静编写,其中第 1、2、5 和 6 章由毛玉萃编写,第 3、4 章由杨文艳编写,第 7 章由秦静编写,最后一章(第 8 章)由邱少明编写。在本书撰写过程中杨春艳、龙翔宇、曾垂军等提供了帮助,在此表示感谢。

  与本书配套的资料包括课件、完成的系统。

  本书的第 2 章节引用了 SQL Server 2008 help、Visual Studio 2010 help 和 MyEclipse 8.6 help 中的部分内容,对原文作者表示感谢。

  由于作者水平有限,书中难免会有不妥之处,敬请广大读者批评指正。

编著者
2014 年 6 月于大连

目录

第1章　系统开发方法简介 …………………………… 1

1.1　原型法简介 ………………………………………… 1
 1.1.1　基本概念 …………………………………… 1
 1.1.2　原型方法的工作流程 ……………………… 1
 1.1.3　原型方法的特点 …………………………… 1
 1.1.4　软件支持环境 ……………………………… 2
 1.1.5　适用范围 …………………………………… 3

1.2　结构化系统开发方法简介 ………………………… 3
 1.2.1　结构化系统开发方法的基本思想 ………… 3
 1.2.2　结构化系统开发的生命周期 ……………… 4
 1.2.3　结构化系统开发方法使用的主要工具 …… 9
 1.2.4　结构化开发方法的特点 …………………… 15
 1.2.5　结构化系统开发方法的优缺点 …………… 16

1.3　面向对象方法简介 ………………………………… 16
 1.3.1　面向对象方法的相关概念 ………………… 17
 1.3.2　面向对象方法的基本思想 ………………… 18
 1.3.3　面向对象方法的开发过程 ………………… 19
 1.3.4　面向对象的建模语言——统一建模语言 … 23
 1.3.5　面向对象方法的特点和面临的问题 ……… 28

第2章　开发技术和环境简介 …………………………… 29

2.1　SQL Server 2008 简介 …………………………… 29
 2.1.1　SQL 简介 …………………………………… 29
 2.1.2　SQL Server 的发展 ………………………… 30
 2.1.3　SQL Server 2008 的版本 …………………… 31
 2.1.4　SQL Server 2008 的新增功能 ……………… 32

         2.1.5　SQL Server 2008 的新增特点 …………………………………… 36
         2.1.6　SQL Server 2008 安装要求 ……………………………………… 37
　2.2　MySQL 简介 ………………………………………………………………… 38
         2.2.1　MySQL ……………………………………………………………… 38
         2.2.2　MySQL 经典应用环境 ……………………………………………… 38
         2.2.3　MySQL 特点 ………………………………………………………… 38
         2.2.4　MySQL 存储引擎 …………………………………………………… 39
         2.2.5　MySQL 应用架构 …………………………………………………… 40
　2.3　Java 及 Java 开发环境简介 ………………………………………………… 41
         2.3.1　Java 起源 …………………………………………………………… 41
         2.3.2　Java 及 Java 平台的组成 …………………………………………… 42
         2.3.3　Java 的版本 ………………………………………………………… 42
         2.3.4　Java 的相关技术和主要特性 ……………………………………… 43
         2.3.5　JSP 简介 …………………………………………………………… 44
         2.3.6　Java 的开发环境 MyEclipse 8.6 简介 …………………………… 46
　2.4　.NET 技术简介 ……………………………………………………………… 46
         2.4.1　.NET 是什么 ………………………………………………………… 46
         2.4.2　.NET 框架 …………………………………………………………… 47
         2.4.3　.NET 的特点 ………………………………………………………… 48
         2.4.4　.NET 的版本 ………………………………………………………… 49
　2.5　三层架构和 MVC 架构简介 ………………………………………………… 49
         2.5.1　三层架构简介 ……………………………………………………… 49
         2.5.2　MVC 框架简介 ……………………………………………………… 51
         2.5.3　三层架构和 MVC 框架的关系 …………………………………… 52

第 3 章　ATM 存取款管理系统设计与实现 ………………………………………… 53
　3.1　项目需求分析 ………………………………………………………………… 53
　3.2　面向对象的分析与设计 ……………………………………………………… 53
         3.2.1　实体类分析与设计 ………………………………………………… 53
         3.2.2　工具类分析与设计 ………………………………………………… 54
         3.2.3　主类分析与设计 …………………………………………………… 54
　3.3　系统实现与测试 ……………………………………………………………… 55
         3.3.1　项目环境准备 ……………………………………………………… 55
         3.3.2　项目类定义与实现 ………………………………………………… 59
         3.3.3　项目测试与改进 …………………………………………………… 64
　3.4　课后训练项目：银行业务调度系统 ………………………………………… 65

第 4 章　Java 在线考试系统设计与实现 ································ 67

4.1　系统分析 ·· 67
4.1.1　需求分析 ·· 67
4.1.2　业务流程分析 ·· 68
4.1.3　数据分析 ·· 69

4.2　系统设计 ·· 69
4.2.1　系统设计思路 ·· 69
4.2.2　功能模块设计 ·· 69
4.2.3　数据库设计 ·· 70
4.2.4　类的分层设计 ·· 72

4.3　系统实现与测试 ·· 78
4.3.1　数据库的建立与连接 ······································ 78
4.3.2　Entity 实体类的实现 ······································ 83
4.3.3　DAO 数据访问类的实现 ···································· 88
4.3.4　GUI 界面类的实现 ·· 98

4.4　项目发布与改进 ·· 110
4.4.1　项目发布 ·· 110
4.4.2　项目改进 ·· 111

第 5 章　网上灯饰店的研究与实现 ·· 113

5.1　网上灯饰店规划 ·· 113
5.1.1　网上商店系统发展和实现网上商店系统的意义 ················ 113
5.1.2　网上灯饰店的需求分析 ···································· 114
5.1.3　网上灯饰店可行性研究 ···································· 115

5.2　网上灯饰店分析 ·· 116
5.2.1　业务流程分析与描述 ······································ 117
5.2.2　数据流程分析与描述 ······································ 118
5.2.3　数据分析 ·· 119

5.3　网上灯饰店设计 ·· 127
5.3.1　网上灯饰店功能结构设计 ·································· 128
5.3.2　网上灯饰店数据库设计 ···································· 129
5.3.3　主要模块功能详细设计 ···································· 134

5.4　网上灯饰店实现 ·· 136
5.4.1　系统运行环境 ·· 136
5.4.2　数据库的建立与连接 ······································ 137
5.4.3　系统实现的总体框架 ······································ 137
5.4.4　数据访问层的设计与实现 ·································· 137

    5.4.5 业务逻辑层的设计与实现 ··· 147
    5.4.6 公共库的设计与实现 ··· 150
    5.4.7 实体模型部分的设计与实现 ··· 154
    5.4.8 部分表示层及控制层的设计与实现 ····································· 158

第6章 家具网站的研究与实现 ·· 183

6.1 系统规划 ··· 183
    6.1.1 系统需求分析和目标设定 ·· 183
    6.1.2 系统可行性分析 ·· 184
6.2 系统分析 ··· 185
    6.2.1 业务流程分析与描述 ·· 185
    6.2.2 数据流程分析与描述 ·· 187
    6.2.3 数据分析与描述 ··· 189
6.3 系统设计 ··· 194
    6.3.1 系统设计思想简介 ··· 195
    6.3.2 系统功能结构设计 ··· 195
    6.3.3 数据库设计 ·· 196
    6.3.4 系统功能详细设计 ··· 199
6.4 系统实现 ··· 203
    6.4.1 数据库的建立与连接 ··· 203
    6.4.2 系统实现总框架简介 ··· 205
    6.4.3 系统实现——DAL 层 ··· 205
    6.4.4 系统实现——USL 层 ··· 219
    6.4.5 系统实现——BLL 层 ··· 252

第7章 网络办公自动化系统的研究与实现 ·· 266

7.1 需求调查分析 ·· 266
    7.1.1 系统定义及可行性分析 ·· 267
    7.1.2 系统需求分析和目标设定 ·· 268
7.2 用例建模 ··· 269
    7.2.1 角色用例图 ·· 269
    7.2.2 模块用例图 ·· 271
7.3 静态建模 ··· 275
    7.3.1 系统类图 ··· 275
    7.3.2 各类之间的关系 ·· 276
7.4 系统设计 ··· 276
    7.4.1 系统功能结构设计 ··· 276
    7.4.2 系统层次结构设计 ··· 276

7.5 动态建模 ··· 277
7.5.1 模块时序图 ··· 277
7.5.2 模块活动图 ··· 278
7.6 输入输出设计 ··· 281
7.6.1 输入设计 ··· 281
7.6.2 输出设计 ··· 281
7.7 物理建模 ··· 281
7.7.1 系统部署 ··· 281
7.7.2 数据库设计 ··· 281
7.7.3 数据库表设计 ··· 282
7.8 系统实现与测试 ··· 284
7.8.1 数据库的建立与连接 ··· 284
7.8.2 系统实现总框架简介 ··· 285
7.8.3 系统实现——MODEL 层 ··· 285
7.8.4 系统实现——DAL 层 ··· 295
7.8.5 系统实现——BLL 层 ··· 317
7.8.6 系统实现——Web 层 ··· 326

第8章 大学毕业(论文)设计管理网站的研究与实现 ··· 335
8.1 用例建模 ··· 337
8.2 静态建模 ··· 337
8.3 系统设计 ··· 340
8.3.1 功能设计 ··· 340
8.3.2 数据库设计 ··· 341
8.4 动态建模 ··· 344
8.5 物理建模 ··· 346
8.6 系统实现与测试 ··· 346
8.6.1 公共部分的设计 ··· 346
8.6.2 学生选题模块的实现 ··· 363
8.6.3 教师确认学生子模块的实现 ··· 368
8.6.4 题目调配子模块的实现 ··· 371
8.6.5 调配教师子模块的实现 ··· 375

参考文献 ··· 377

第 1 章 系统开发方法简介

20 世纪 70 年代以来,在西方经历了"软件危机"以后,开发研究人员开始重视系统开发方法的研究,提出了许多新的系统开发方法。目前,常用的开发方法有原型法、结构化系统开发方法和面向对象法等。

1.1 原型法简介

1.1.1 基本概念

原型(prototype)是指由系统分析设计人员与用户合作,在短期内定义用户基本需求的基础上,开发出来的只具备基本功能、实验性的、简易的应用软件。原型方法舍弃了一步步周密细致地调查分析,然后逐步整理出文字档案,最后才能让用户看到结果的烦琐作法。

原型方法(prototyping)是 20 世纪 80 年代随着计算机软件技术的发展,特别是在关系数据库系统(RDBS)、第四代程序生成语言和各种系统开发生成环境的基础上,提出的一种从设计思想、工具到手段都全新的系统开发方法。

原型法一开始就凭借着系统开发人员对用户要求的理解,在强有力的软件环境支持下,给出一个实实在在的系统原型,然后与用户反复协商修改,最终形成实际系统。

1.1.2 原型方法的工作流程

原型方法的工作流程如图 1.1 所示。首先用户提出开发要求,开发人员识别和归纳用户要求,根据识别、归纳的结果,构造出一个原型(即程序模块),然后同用户一道评价这个原型。如果根本不行,则回到第二步重新归纳问题、构造原型;如果不满意,则修改原型,直到用户满意为止。这就是原型法工作的一般流程。

1.1.3 原型方法的特点

原型方法无论从原理到流程都是十分简单的,并无任何高深的理论和技术,能在实践

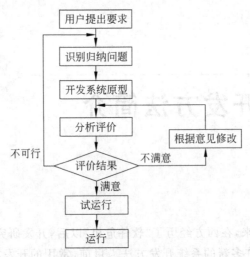

图 1.1　原型方法的工作流程

中获得巨大成功主要在于原型方法具有如下几方面的特点。

(1) 从认识论的角度来看,原型方法更多地遵循了人们认识事物的规律,因而更容易为人们所普遍接受,这主要表现在:

① 人们认识任何事物都不可能一次就完全了解,并把工作做得尽善尽美。

② 认识和学习的过程都是循序渐进的。

③ 人们对于事物的描述,往往都是受环境的启发而不断完善的。

④ 人们批评指责一个已有的事物,要比空洞地描述自己的设想容易得多,改进一些事物要比创造一些事物容易得多。

(2) 原型方法将模拟的手段引入系统分析的初期阶段,沟通了人们的思想,缩短了用户和系统分析人员之间的距离,解决了结构化方法中最难于解决的一环。这主要表现在:

① 所有问题的讨论都是围绕某一个确定原型而进行的,彼此之间不存在误解和答非所问的可能性,为准确认识问题创造了条件。

② 有了原型后才能启发人们对原来想不起来或不易准确描述的问题有一个比较确切的描述。

③ 能够及早地暴露出系统实现后存在的一些问题,促使人们在系统实现之前就加以解决。

(3) 充分利用了最新的软件工具,摆脱了老一套的工作方法,从而使系统开发的时间、费用大大减少,效率、技术等方面都极大提高了。

1.1.4　软件支持环境

原型方法有很多长处,有很大的推广价值。但必须指出,它的推广应用必须要有一个强有力的软件支持环境作为背景,没有这个背景它将变得毫无价值。一般认为原型方法所需要的软件支撑环境主要有:

(1) 一个方便灵活的关系数据库系统(RDBS)。

（2）一个与 RDBS 相对应的、方便灵活的数据字典，它具有存储所有实体的功能。

（3）一套与 RDBS 相对应的快速查询系统，能支持任意非过程化的（即交互定义方式）组合条件查询。

（4）一套高级的软件工具（如 4GL 或信息系统开发生成环境等）用以支持结构化程序，并且允许采用交互的方式迅速进行书写和维护，可产生任意程序语言的模块（即原型）。

（5）一个非过程化的报告或屏幕生成器，允许设计人员详细定义报告或屏幕输出样本。

1.1.5 适用范围

作为一种具体的开发方法，原型法不是万能的，有其一定的适用范围和局限性。这主要表现在：

（1）对于一个大型的系统，如果不经过系统分析就进行整体性划分，想要直接用屏幕来一个一个地模拟是很困难的。

（2）对于大量运算的、逻辑性较强的程序模块，原型方法很难构造出模型来供人评价。因为这类问题没有那么多的交互方式（如果有现成的数据或逻辑计算软件包，则情况例外），也不是三言两语就可以把问题说清楚的。

（3）对于原基础管理不善、信息处理过程混乱的问题，使用有一定的困难。首先由于工作过程不清，构造原型有一定困难；其次由于基础管理不好，没有科学合理的方法可依，系统开发容易走上机械地模拟原来手工系统的轨道。

（4）对于一个批处理系统，其大部分是内部处理过程，这时用原型方法有一定的困难。

原型方法是在系统开发中的一种简单的模拟方法，是人类认识系统开发规律道路上的"否定之否定"。它站在前者的基础之上，借助于新一代的软件工具，螺旋式地上升到了一个新的更高的起点，它"扬弃"了结构化系统开发方法的某些烦琐细节，继承了其合理的内核，是对结构化开发方法的发展和补充。

1.2 结构化系统开发方法简介

1.2.1 结构化系统开发方法的基本思想

结构化系统开发方法（structured system development methodologies）是开法方法中应用最普遍、最成熟的一种。

结构化系统开发方法的基本思想是用系统工程的思想和工程化的方法，按用户至上的原则，结构化、模块化、自顶向下地对系统进行分析和设计，在实施过程中自底向上逐步实现整个系统。

1.2.2 结构化系统开发的生命周期

用结构化系统开发方法开发一个系统时,将整个开发过程划分为五个首尾相连接的阶段,一般称为系统开发的生命周期(life cycle),如图 1.2 所示。

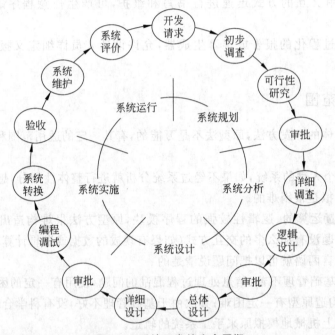

图 1.2 系统开发的生命周期

1. 系统规划阶段

系统规划阶段的主要任务是根据用户的系统开发请求,进行初步调查,明确问题,确定系统目标和系统主要功能结构,确定分阶段实施进度,然后进行可行性研究。

系统规划的步骤是:规划基本问题的确定;收集信息;现状的评价和约束的识别;设置具体目标和初步方案;规划内容及其相关性分析;目标的分析及实现的优先级;人员组织;实施进度计划,包括经费预算和使用计划;成本分析、效益初步分析;开发平台软硬件环境(不一定马上购买);可行性分析。

系统规划的成果包括:

(1) 可行性报告及其审批;

(2) 开发团队的组织。

可行性报告中包括系统总体方案和对方案进行的可行性分析。

① 总体方案内容主要有:

- 确定系统目标;
- 设计系统主要功能结构;
- 系统开发的初步计划;
- 投资回报时间表。

② 可行性分析主要方面有：
- 经济可行性分析；
- 技术可行性；
- 管理可行性。

2. 系统分析阶段

系统分析阶段的任务是：分析业务流程，分析数据与数据流程，分析功能与数据之间的关系，最后提出分析处理方式和新系统逻辑方案。

系统分析阶段由详细调查和逻辑设计两部分组成。

在详细调查中进行的工作包括：
- 确定调查方法；
- 组织结构调查和描述；
- 业务流程调查和描述；
- 信息调查和描述。

在逻辑设计中提出系统逻辑模型。在系统逻辑模型中主要包括：
- 功能模型，描述新系统的功能；
- 数据流程图，描述信息和信息的流动；
- 信息模型，采用数据字典和数据库结构等方式描述信息。

3. 系统设计阶段

系统设计阶段的任务是：总体结构设计，代码设计，数据库/文件设计，输入输出设计，模块结构与功能设计。与此同时根据总体设计的要求购置与安装一些设备，进行试验，最终给出系统设计方案。

这一阶段分为总体设计和详细设计两个步骤。

1）总体结构的设计

在总体结构的设计步骤中完成系统的总体结构的设计，并用模块结构图表示出来。

进行总体结构设计的原则有如下六点：

（1）降低模块的耦合性，提高模块的内聚性。
（2）保持适中的模块规模。
（3）模块应具有高扇入和适当的扇出。
（4）软件结构中的深度和宽度不宜过大。
（5）模块的作用域应处于其控制域范围之内。
（6）尽量降低模块的接口复杂度。

进行总体设计主要采用面向数据流的体系设计过程。运用面向数据流的方法进行总体结构的设计时，应该首先对系统分析阶段得到的数据流图进行复查，必要时进行修改和精化；接着在仔细分析系统数据流图的基础上，确定数据流图的类型，并按照相应的设计步骤将数据流图转化为系统总体结构，并用模块结构图予以描述；最后还要根据系统结构设计的原则对得到的总体结构进行优化和改进。

2）详细设计

详细设计主要包括如下内容：

（1）数据库/数据文件设计。

（2）运行设计，用于说明软件的运行模块组合、运行控制方式及运行时间等。

（3）计算机网络设计。

（4）出错处理设计。

（5）代码设计。

（6）用户界面设计，也称输入输出设计。

（7）计算机处理。

数据库设计（database design）是指根据用户的需求，在某一具体的数据库管理系统上，设计数据库的结构和建立数据库的过程。数据库设计是一种"反复探寻，逐步求精"的过程，也就是规划和结构化数据库中的数据对象以及这些数据对象之间的关系的过程。

数据库设计的步骤如下。

① 需求分析：调查和分析用户的业务活动和数据的使用情况，弄清所用数据的种类、范围、数量以及它们在业务活动中交流的情况，确定用户对数据库系统的使用要求和各种约束条件等，形成用户需求规约。

② 概念设计：对用户要求描述的现实世界，通过对其中诸处的分类、聚集和概括，建立抽象的概念数据模型。概念模型使用实体-联系模型（E-R 模型）表示。

③ 逻辑设计：主要工作是将现实世界的概念数据模型设计成数据库的一种逻辑模式，即适应于某种特定数据库管理系统所支持的逻辑数据模式。逻辑数据模式主要有层次模型、网络模型和关系模型。

④ 物理设计：根据特定数据库管理系统所提供的多种存储结构和存取方法等依赖于具体计算机结构的各项物理设计措施，对具体的应用任务选定最合适的物理存储结构（包括文件类型、索引结构和数据的存放次序与位逻辑等）、存取方法和存取路径等。这一步设计的结果就是所谓"物理数据库"。

计算机网络结构设计的主要内容有：

① 网络拓扑设计。

② 两层结构和三层结构。

③ 网络协议。

④ 有线和无线网络的选择和连接。

⑤ 网络设备选型。

⑥ 内部网路如何接入因特网。

系统出错是不可避免的，为了使系统在出现错误的情况下，尽量降低由于错误造成的损失，在设计系统时必须设计必要的出错处理。

出错处理设计主要包括：

① 可能出错的情况、出错类型和位置。

② 出错信息的设计：采用错误提示窗口向用户提示错误，并友好地处理错误。例如，用户登录失败时，根据失败原因进行提示；用户输入不正确时，进行适当提示。

③ 补救措施，说明故障出现后可能采取的变通措施，主要包括：
- 对于软错误，需要在添加/修改操作中及时对输入数据进行验证，分析错误的类型，并且给出相应的错误提示语句，传送到客户端的浏览器上；
- 对于硬错误，在可能出错的地方中输出相应的出错语句，并将程序重置，最后返回输入阶段；
- 说明准备采用的后备技术，即当原始数据丢失时启用的副本的建立和启动技术，例如周期性地把磁盘信息记录在案；
- 恢复及再启动技术说明将使用的恢复再启动技术，是软件从故障点恢复执行或是软件从头开始重新运行的方法。

代码就是处理对象的代号或标识符号。代码设计就是设计和构造一套代码生成的规则和方法。代码设计的一般原则如下：

① 符合现有标准，其选择顺序一般是国际标准、国标、行业标准、企业标准；
② 具有唯一性；
③ 直观、逻辑性强、短小、便于记忆；
④ 具有可扩充性。

用户界面设计是一个复杂的有不同学科参与的工程，认知心理学、设计学、语言学等在此都扮演着重要的角色。用户界面设计的三大原则是：置界面于用户的控制之下；减少用户的记忆负担；保持界面的一致性。一个设计良好的用户界面可以提高工作效率，使用户从中获得乐趣，减少由于界面问题而造成错误。

计算机处理就是针对模块结构图中的由计算机处理的每个模块确定具体算法并选择某种表达工具将算法的详细处理过程描述出来。同时确定模块接口的具体细节，并为每个模块设计一组测试用例。

3) 系统设计阶段的成果

主要成果是系统设计说明书文档。该文档应包括以下主要内容：

(1) 模块结构图及每一模块详细说明；
(2) 数据库设计说明；
(3) 计算机网络系统设计说明；
(4) 代码设计说明；
(5) 用户界面设计说明；
(6) 计算机处理过程说明；
(7) 实施费用估计。

4．系统实施阶段

系统实施阶段的任务是：同时进行编程（由程序员执行）和人员培训（由系统分析设计人员培训业务人员和操作员），以及数据准备（由业务人员完成），然后投入试运行。

系统实现阶段的任务是根据系统设计的结果，完成程序设计、调试、系统切换以及为了使系统有效运行的需要做的其他一系列工作。

系统实现应做的准备工作包括：

(1) 制定实现计划。

(2) 硬件设备及系统软件购置、安装和调试。
(3) 数据的准备。
(4) 人员的培训。

程序设计是这一阶段的主要工作,也关系着系统开发的成败。进行程序设计首先要为待开发项目选择合适的程序设计语言,应充分考虑到项目的各种需求,结合各种语言的心理特性、工程特性、技术特性以及应用特点,尽量选取实现效率高且易于理解和维护的语言。由于程序设计语言的选择往往会受到各种实际因素的制约和限制,因此选择语言时不能只考虑理论上的标准,而是要同时兼顾理论标准和实用标准。

① 理论标准包括:
- 理想的模块化机制、易于阅读和使用的控制结构及数据结构;
- 完善、独立的编译机制。

② 实用标准包括:
- 系统用户的要求;
- 工程的规模;
- 软件的运行环境;
- 可以得到的软件开发工具;
- 软件开发人员的知识面,掌握几种程序设计语言;
- 软件的可移植性要求;
- 软件的应用领域。

为了提高编程效果,应养成良好的编码风格。编码风格是指在不影响程序正确性和效率的前提下,有效编排和合理组织程序的基本原则。一个具有良好编码风格的程序主要表现为可读性好、易测试、易维护。由于测试和维护阶段的费用在软件开发总成本中所占比例很大,因此编码风格的好坏直接影响整个软件开发中的成本耗费。特别是在需要团队合作开发大型软件的时候,编码风格显得尤为重要。若团队中的成员不注重自己的编码风格,则会严重影响与其他成员的合作和沟通,最终将可能导致软件质量上出现问题。

系统测试是将经过集成测试的软件,作为计算机系统的一个部分,与系统中其他部分结合起来,在实际运行环境下对计算机系统进行一系列严格有效的测试,以发现软件潜在的问题,保证系统的正常运行。系统测试的目的是验证最终软件系统是否满足用户规定的需求。系统测试方法主要有黑盒法、白盒法、α测试和β测试。

系统测试过程如下:
① 制定系统测试计划;
② 设计系统测试用例;
③ 执行系统测试;
④ 缺陷管理与改错。

系统切换是指系统开发完成后新旧系统之间转换。主要的方法有直接切换、并行切换和分段切换。

5. 系统运行阶段

系统运行阶段的任务是同时进行系统的日常运行管理、评价、监理审计三部分工作，然后分析运行结果。如果运行结果良好，则送管理部门，指导生产经营活动；如果有点问题，则要对系统进行修改、维护或者是局部调整；如果出现了不可调和的大问题，则用户将会进一步提出开发新系统的要求，这标志着旧系统生命的结束，新系统的诞生。这全过程就是系统开发生命周期。在每一阶段均有小循环，在不满足要求时，修改或返回到起点。

1.2.3 结构化系统开发方法使用的主要工具

在结构化系统开发方法中，提供了一套标准的工具帮助研究开发人员进行研究和表达研究结果。这些工具主要是表和图。主要的工具列于表1.1中。

表1.1 结构化系统开发方法使用的主要工具

序号	工具名称	作用	应用阶段
1	业务流程图	业务流程分析	系统分析
2	数据流图（DFD）	数据流程分析	系统分析、系统设计
3	数据字典（DD）	数据定义	系统分析
4	决策树、判定表	处理功能描述	系统分析、系统设计
5	E-R图（ERD）	数据库设计	系统设计
6	U/C矩阵	系统结构设计	系统设计
7	模块结构图	系统结构和模块设计	系统设计
8	网络拓扑图	网络图拓扑结构描述	系统设计
9	程序流程图	算法描述	系统设计
10	N-S图	算法描述	系统设计
11	系统配置图	系统硬软件配置	系统实施

下面简单介绍几个主要的工具。

1. 业务流程图

业务流程图是一种描述系统内各单位、人员之间业务关系、作业顺序和信息流向的图，利用它可以帮助分析人员描述系统组织结构、业务流程，找出业务流程中的不合理流向。业务流程图描述的是完整的业务流程，以业务处理过程为中心。业务流程图是一种系统分析人员和业务人员都懂的共同语言。

业务流程图的绘制是按照业务的实际处理步骤和过程进行的，主要分为两步：

（1）现行系统业务流程总结。在画业务流程图之前，要对现行系统进行详细调查，并写出现行系统业务流程总结。

（2）业务流程图的绘制。

业务流程图中的基本符号及含义，如图1.3所示。

图1.3 业务流程图中的基本符号及含义

业务流程图特点如下:
(1) 图的形式是按业务部门划分的横式图。
(2) 图描述的主体是数据或单据的业务处理。
(3) 数据或单据流动路线与实际业务处理过程一一对应。
(4) 图中数据或单据是有"生"、"死"的,即用它的一次生命周期来表示出一笔业务的处理情况。

业务流程图作用如下:
(1) 制作流程图的过程是全面了解业务处理的过程,是进行系统分析的依据。
(2) 它是系统分析员、管理人员、业务操作人员相互交流思想的工具。
(3) 系统分析员可直接在业务流程图上拟出可以实现计算机处理的部分。
(4) 用它可分析出业务流程的合理性。

2. 数据流图

数据流图(Data Flow Diagram,DFD)是结构化系统开发方法中使用的工具,从数据传递和加工角度,以图形方式表达系统的逻辑功能、数据在系统内部的逻辑流向和逻辑变换过程,是表示系统逻辑模型的一种图示方法。

数据流程图中有以下几种主要元素(见图1.4)。

图1.4 数据流图中的元素

数据流:数据流是数据在系统内传播的路径,因此由一组成分固定的数据组成,如订票单由旅客姓名、年龄、单位、身份证号、日期、目的地等数据项组成。由于数据流是流动中的数据,所以必须有流向,除了与数据存储之间的数据流不用命名外,数据流应该用名词或名词短语命名。

数据源(终点):代表系统之外的实体,可以是人、物或其他软件系统。

数据处理:处理或加工是对数据进行处理的单元,它接收一定的数据输入,对其进行处理,并产生输出。

数据存储:表示信息的静态存储,可以代表文件、文件的一部分、数据库的元素等。

数据流图有两种典型结构,一是变换型结构,它所描述的工作可表示为输入、主处理和输出,呈线性状态,如图1.5(a)所示。另一种是事务型结构,这种数据流图呈束状,如

图1.5(b)所示,即一束数据流平行流入或流出,可能同时有几个事务要求处理。

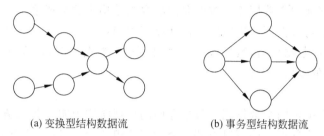

(a) 变换型结构数据流　　　　(b) 事务型结构数据流

图 1.5　数据流图的两种结构

在绘制单张数据流图时,必须注意以下原则:

(1) 一个加工的输出数据流不应与输入数据流同名,即使它们的组成成分相同。

(2) 保持数据守恒。也就是说,一个加工所有输出数据流中的数据必须能从该加工的输入数据流中直接获得,或者说是通过该加工能产生的数据。

(3) 每个加工必须既有输入数据流,又有输出数据流。

(4) 所有的数据流必须以一个外部实体开始,并以一个外部实体结束。

(5) 外部实体之间不应该存在数据流。

数据流图绘制步骤:

(1) 确定系统的输入输出。

由于系统究竟包括哪些功能可能一时难于弄清楚,所以可使范围尽量大一些,把可能有的内容全部都包括进去。此时,应该向用户了解"系统从外界接收什么数据"、"系统向外界送出什么数据"等信息,然后,根据用户的答复画出数据流图的外围。

(2) 由外向里画系统的顶层数据流图。

首先,将系统的输入数据和输出数据用一连串的加工连接起来。在数据流的值发生变化的地方就是一个加工。接着,给各个加工命名。然后,给加工之间的数据命名。最后,给文件命名。

(3) 自顶向下逐层分解,绘出分层数据流图。

对于大型的系统,为了控制复杂性,便于理解,需要采用自顶向下逐层分解的方法进行,即用分层的方法将一个数据流图分解成几个数据流图分别表示。

3. 数据字典

数据字典(Data Dictionary,DD)是指对数据的数据项、数据结构、数据流、数据存储、处理逻辑、外部实体等进行定义和描述,其目的是对数据流程图中的各个元素做出详细说明。数据字典则是系统中各类数据描述的集合,是进行详细的数据收集和数据分析所获得的主要成果。

数据字典通常包括数据项、数据结构、数据流、数据存储和处理过程五部分。其中数据项是数据的最小组成单位,若干个数据项可以组成一个数据结构,数据字典通过对数据项和数据结构的定义来描述数据流、数据存储等的逻辑内容。

数据字典各部分的描述如下。

1) 数据项

数据项是数据流图中数据流或数据存储的数据结构中的数据项说明。数据项是不可再分的数据单位。对数据项的描述通常包括以下内容：

数据项描述={数据项名,数据项含义说明,别名,数据类型,长度,取值范围,取值含义,与其他数据项的逻辑关系}

2) 数据结构

数据结构是数据流图中数据流或数据存储的数据结构说明。数据结构反映了数据之间的组合关系。一个数据结构可以由若干个数据项组成,也可以由若干个数据结构组成,或由若干个数据项和数据结构混合组成。对数据结构的描述通常包括以下内容：

数据结构描述={数据结构名,含义说明,组成:{数据项或数据结构}}

3) 数据流

数据流是数据流图中流线的说明。数据流是数据结构在系统内传输的路径。对数据流的描述通常包括以下内容：

数据流描述={数据流名,说明,数据流来源,数据流去向,组成:{数据结构},平均流量,高峰期流量}

4) 数据存储

数据存储是数据流图中数据存储的存储特性说明。数据存储是数据结构停留或保存的地方,也是数据流的来源和去向之一。对数据存储的描述通常包括以下内容：

数据存储描述={数据存储名,说明,编号,流入的数据流,流出的数据流,组成:{数据结构},数据量,存取方式}

5) 处理过程

处理过程是数据流图中处理或加工的说明。数据字典中只需要描述处理过程的说明性信息,通常包括以下内容：

处理过程描述={处理过程名,说明,输入:{数据流},输出:{数据流},处理:{简要说明}}

数据字典也是在数据库设计时用到的一种工具,用来描述数据库中基本表的设计,主要包括字段名、数据类型、主键、外键等描述表的属性的内容。

6) 外部实体

外部实体是数据流图中外部实体的说明。数据字典中只需要描述外部实体编号、名称和说明即可。

4. 决策树、判定表

1) 决策树

每个决策或事件都可能引出两个或多个事件,从而导致不同的结果。把这种决策分支画成图形很像一棵树的枝干,故称决策树(decision tree)。决策树一般都是自上而下生成的。决策树就是将决策过程中各个阶段之间的结构绘制成一张树状图。

一个简单的决策树如图 1.6 所示。

决策树代表的意义：左边是树根,是决策序列的条件取值状态;右边是树叶,表示应该采取的动作。

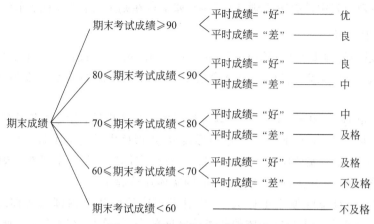

图 1.6 期末成绩评定的决策树

决策树可以是二叉的,也可以是多叉的。

2) 判断表

判断表又称决策表,是一种呈表格状的表达工具,适用于描述处理判断条件较多,各条件又相互组合、有多种决策方案的情况。决策表能精确而简洁地描述复杂逻辑的方式,将多个条件与这些条件满足后要执行的动作相对应。决策表能将多个独立的条件和多个动作直接的联系清晰地表示出来。

一个简单的判断表如表 1.2 所示。

表 1.2 赊购处理的判定表

条件和结果 \ 不同条件组合	1	2	3	4	5	6	7	8
C1:购货金额≥500	Y	Y	Y	Y	N	N	N	N
C2:过去交款情况好	Y	Y	N	N	Y	Y	N	N
C3:库存货物够	Y	N	Y	N	Y	N	Y	N
A1:立即发货	X				X		X	
A2:要求先付款			X					
A3:暂挂,货到后发货		X				X		X
A4:退回				X				

判断表的绘制步骤如下:

(1)确定判断要采用的相关因素,即决策中的必要条件,而这些条件的选择必须是发生或不发生两种值。

(2)在各种不同的条件下确定相应的行动。

(3)排列出各种不同条件之间的所有组合,Y 和 N 分别表示发生和不发生。

(4)确定在不同组合下应选择的行动,即形成条件项和行动项相关联系的决策规则,以这些规则指导决策。

(5) 检验判断表中的决策规则是否冗余,如果存在则进行合并,消除冗余。

5. 实体-联系图

实体-联系图简称 E-R 图(Entity Relationship Diagram),提供了表示实体类型、属性和联系的方法,用来描述现实世界的概念模型。构成 E-R 图的基本要素是实体型、属性和联系。

实体型(entity):相同属性的实体具有相同的特征和性质,用实体名及其属性名集合来抽象和刻画同类实体;在 E-R 图中用矩形表示,矩形框内写明实体名。

属性(attribute):实体所具有的某一特性,一个实体可由若干个属性来刻画。在 E-R 图中用椭圆形表示,并用无向边将其与相应的实体连接起来。

联系(relationship):联系也称关系,信息世界中反映实体内部或实体之间的联系。实体内部的联系通常是指组成实体的各属性之间的联系;实体之间的联系通常是指不同实体集之间的联系。在 E-R 图中用菱形表示,菱形框内写明联系名,并用无向边分别与有关实体连接起来,同时在无向边旁标上联系的类型。联系可分为以下三种类型:

(1) 一对一联系(1:1);
(2) 一对多联系(1:N);
(3) 多对多联系(M:N)。

E-R 图的绘制步骤如下:
(1) 确定所有的实体集合;
(2) 选择实体集应包含的属性;
(3) 确定实体集之间的联系;
(4) 确定实体集的关键字,用下划线在属性上表明关键字的属性组合;
(5) 确定联系的类型,在用线将表示联系的菱形框联系到实体集时,在线旁注明是 1 或 N(多)来表示联系的类型。

一个简单的 E-R 图如图 1.7 所示。

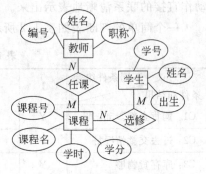

图 1.7 一个简单的 E-R 图

6. 模块结构图

模块结构图是由 Yuordon 在 1974 年提出的,用于描述软件系统的组成结构及其相互关系,它反映了系统的结构(即模块划分),同时也反映了模块之间的关系;模块结构图中最基本的元素是模块。模块划分应体现出信息隐蔽、高内聚、低耦合的特性。模块是指具有一定功能的相对独立的单元。在一个系统中模块是分层次的;一个模块本身具有三个基本属性,一是功能,说明模块实现什么,一般以模块名的形式给出;二是逻辑,描述模块内部如何实现所要求的功能,一般用功能处理表达工具给予描述,常用的处理表达工具有决策树、判断表、伪码语言和 Warnier-Orr 图;三是状态,描述模块的使用环境、条件和模块间的相互关系,一般用特定的工具和文字进行说明。

总体设计可以采用面向数据流的设计,也可采用面向事务的设计,还可以采用面向数据流设计和面向事务设计相结合。根据数据流图"导出"模块结构图。如果采用面向数据流的设计,就根据数据流采用自顶向下、逐步求精的设计方法,按照系统的层次结构进行逐步分解,并以分层数据处理过程反应。

图 1.8 是一个简单的模块结构图,基本图像符号有模块、调用和模块之间的通信。

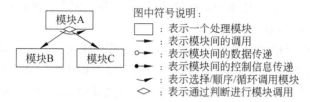

图 1.8　简单的模块结构图

7．程序流程图

程序流程图是最早出现且使用较为广泛的算法表达工具之一,能够有效地描述问题求解过程中的程序逻辑结构。程序流程图中经常使用的基本符号如图 1.9 所示。

图 1.9　程序流程图中经常使用的基本符号

8．N-S 图

N-S 图又称为盒图,它是为了保证结构化程序设计而由 Nassi 和 Shneiderman 共同提出的一种图形工具。在 N-S 图中,所有的程序结构均使用矩形框表示,它可以清晰地表达结构中的嵌套及模块的层次关系。N-S 图中,基本控制结构的表示符号如图 1.10 所示。

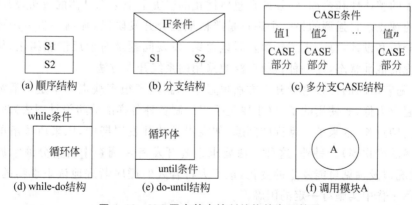

图 1.10　N-S 图中基本控制结构的表示符号

1.2.4　结构化开发方法的特点

结构化系统开发方法具有如下主要特点。

(1) 用户至上。用户对系统开发的成败是至关重要的,故在系统开发过程中要面向用户,让用户参与系统开发的全过程,要充分了解用户的需求和愿望。

(2)"自顶向下"整体性的分析与设计和"自底向上"逐步实施的系统开发过程。在系统分析与设计时从整体、全局考虑,采用自顶向下的工作原则;系统实现时,则要根据设计的要求先编制和测试一个个具体的功能模块,然后按照系统设计的结构,将这些模块一个个拼接到一起进行调试,从而构成完整系统。

(3)把开发过程严格区分为几个工作阶段。把整个系统开发过程划分为若干个工作阶段,每个阶段都有其明确的任务和目标,各阶段次序不能改变。在实际开发过程中要求严格按照划分的工作阶段,一步步地展开工作,如遇到较小、较简单的问题,可跳过某些步骤,但不可打乱或颠倒。

(4)开发过程工程化,文档资料标准化。要求开发过程的每一步都按工程标准规范化,文档资料按照标准化的要求进行撰写和归档管理。

(5)深入调查研究。强调在系统设计之前,深入实际单位,详细地调查研究,弄清实际业务处理过程中的每一个环节,然后进行分析研究,制定出科学合理的符合用户需求的新系统设计方案。

(6)充分预料可能发生的变化。系统开发是一项耗费人力、财力、物力且周期很长的工作,一旦周围环境(如组织的内外部环境、信息处理模式、用户需求等)发生变化,都会直接影响到系统的开发工作,所以结构化开发方法强调在系统调查和分析时对将来可能发生的变化给予充分的重视,强调所设计的系统对环境的变化具有一定的适应能力。

1.2.5 结构化系统开发方法的优缺点

结构化系统开发方法是在对传统的、自发的系统开发方法批判的基础上,通过很多学者的不断探索和努力而建立起来的一种系统化方法。这种方法的突出优点就是它强调系统开发过程的整体性和全局性,强调在整体优化的前提下来考虑具体的分析设计问题,即自顶向下的观点。它强调的另一个观点是严格地区分开发阶段,强调一步一步地严格地进行系统分析和设计,每一步工作都及时地总结,发现问题及时地反馈和纠正,从而避免了开发过程的混乱状态,是一种目前广泛被采用的系统开发方法。

但是,随着时间的推移这种开发方法也逐渐暴露出了很多缺点和不足。最突出的表现是它的起点太低,所使用的工具(主要是手工绘制各种各样的分析设计图表)落后,致使系统开发周期过长,带来了一系列的问题(如在漫长的开发周期中,原来所了解的情况可能发生较多的变化等)。另外,这种方法要求系统开发者在调查中就充分地掌握用户需求、管理状况以及预见可能发生的变化,这不大符合人们循序渐进地认识事物的规律性,因此在实际工作中实施有一定的困难。

1.3 面向对象方法简介

面向对象系统的开发方法是从 20 世纪 80 年代各种面向对象的程序设计方法(如 small talk,C++等)逐步发展而来的。面向对象方法(Object Oriented,简称 OO 方法)一反那种功能分解方法只能单纯反映管理功能的结构状态,数据流程模型(Data Flow Diagram,DFD)只侧重反映事物的信息特征和流程,信息模拟只能被动地迎合实际问题需要的做法,从面向对象的角度为我们认识事物,进而为开发系统提供了一种全新的方法。

1.3.1 面向对象方法的相关概念

面向对象的基本概念主要有面向对象、对象(模块、继承和类比、动态连接)、类(包括抽象、继承、封装、重载、多态)、消息和结构。

1. 面向对象

对于这一概念至今还没有统一的定义,一般定义为:按人们认识客观世界的系统思维方式,采用基于对象(实体)的概念建立模型,模拟客观世界分析、设计、实现软件的办法。通过面向对象的理念使计算机软件系统能与现实世界中的系统一一对应。

面向对象有抽象性、封装性、继承性、多态性等优异特性。

2. 对象

对象(object)指现实世界中各种各样的实体。它可以指具体的事物也可以指抽象的事物。每个对象都有自己的内部状态和运动规律。在面向对象概念中把对象的内部状态称为属性、运动规律称为方法或事件。对象是一些属性和专用服务(一组操作)的封装体,它是问题空间中一些事务的抽象。属性的值刻画了一个对象的状态,操作是对象的行为,通过操作改变对象的状态。对象是面向对象方法的主体,对象有以下特征。

(1) 模块性(modularity):对象是一个独立存在的实体,从外部可以了解它的功能,但其内部细节是"隐蔽"的,它不受外界干扰。对象之间的相互依赖性很小,因而可以独立地被其他各个系统所选用。

(2) 继承(inheritance)和类比性(ahalogy):事物之间都有一定的相互联系,事物在整体结构中都会占有它自身的位置。在对象之间有属性关系的共同性,在OO方法学中称之为继承性,即子模块继承了父模块的属性。通过类比方法抽象出典型对象的过程称为类比。

(3) 动态连接性:各种对象之间统一、方便、动态的消息传送机制。

3. 类

类(class)是具有相似属性和行为的对象的集合(或称抽象)。类的概念来自于人们认识自然、认识社会的过程。在这一程中,人们主要使用两种方法:由特殊到一般的归纳法和由一般到特殊的演绎法。对于一个具体的类,它有许多具体的个体,这些个体称为"对象"。类的属性是指类集合中对象的共同属性;类的行为是指类集合中对象的共同行为。类具有抽象、继承、封装、多态和重载五个基本特性。

(1) 抽象(abstractrnction):类的定义中明确指出类是一组具有内部状态和运动规律对象的抽象,抽象是一种从一般的观点看待事物的方法,它要求我们集中于事物的本质特征(内部状态和运动规律),而非具体细节或具体实现。面向对象鼓励我们用抽象的观点来看待现实世界,也就是说,现实世界是一组抽象的对象——类组成的。

(2) 继承：继承是类不同抽象级别之间的关系。类的定义主要有归纳和演绎两种办法；由一些特殊类归纳出来的一般类称为这些特殊类的父类，特殊类称为一般类的子类，同样父类可演绎出子类；父类是子类更高级别的抽象。子类可以继承父类的所有内部状态和运动规律。在计算机软件开发中采用继承性，提供了类的规范的等级结构；通过类的继承关系，使公共的特性能够共享，提高了软件的重用性。

(3) 封装(encapsulation)：对象间的相互联系和相互作用过程主要通过消息机制得以实现。对象之间并不需要过多地了解对方内部的具体状态或运动规律。面向对象的类是封装良好的模块，类定义将其说明(用户可见的外部接口)与实现(用户不可见的内部实现)显式地分开，其内部实现按其具体定义的作用域提供保护。类是封装的最基本单位。封装防止了程序相互依赖性而带来的变动影响。在类中定义的接收对方消息的方法称为类的接口。

(4) 多态(ploymorphism)(覆盖)：多态性是指同名的方法可在不同的类中具有不同的运动规律。在父类演绎为子类时，类的运动规律也同样可以演绎，演绎使子类的同名运动规律或运动形式更具体，甚至子类可以有不同于父类的运动规律或运动形式。不同的子类可以演绎出不同的运动规律。

(5) 重载(overload)：重载指类的同名方法在给其传递不同的参数是可以有不同的运动规律。在对象间相互作用时，即使接收消息对象采用相同的接收办法，但消息内容的详细程度不同，接收消息对象内部的运动规律也可能不同。

4. 消息

消息(message)是指对象间相互联系和相互作用的方式。一个消息主要由五部分组成，即发送消息的对象、接收消息的对象、消息传递办法、消息内容(参数)、反馈。一条消息告诉一个对象做什么，它指出：发送者、接收者、需要执行的服务、需要的参数。

5. 包(或称结构)

现实世界中不同对象间的相互联系和相互作用构成了各种不同的系统，不同系统间的相互联系和相互作用构成了更庞大的系统，进而构成了整个世界。在面向对象概念中把这些系统称为包(package)。

6. 包的接口类

在系统间相互作用时为了蕴藏系统内部的具体实现，系统通过设立接口界面类或对象来与其他系统进行交互；让其他系统只看到是这个接口界面类或对象，这个类在面向对象中称为接口类(interface)。

1.3.2 面向对象方法的基本思想

面向对象方法认为，客观世界是由各种各样的对象组成的，每种对象都有各自的内部状态和运动规律，不同的对象之间的相互作用和联系就构成了各种不同的系统。当我们设计和实现一个客观系统时，如能在满足需求的条件下，把系统设计成由一些不可变的(相对固定)部分组成的最小集合，这个设计就是最好的。而这些不可变的部分就是所谓的对象。

因此，以对象为主体的面向对象方法的基本思想可以归纳为：

（1）客观事物都是由对象组成的，对象是在原事物基础上抽象的结果。任何复杂的事物都可以通过对象的某种组合结构构成。

（2）对象由属性和方法组成。属性反映了对象的信息特征，如特点、值、状态等。而方法（method）则是用来定义改变属性状态的各种操作。

（3）对象之间的联系主要是通过传递消息实现的，传递的方式是通过消息模式和方法所定义的操作过程来完成的。

（4）对象可按其属性进行归类。类有一定的结构，类上可以有超类，类下可以有子类。这种对象或类之间的层次结构是靠继承关系维系的。

（5）对象是一个被严格模块化了的实体，称为封装。这种封装的对象满足软件工程的一切要求，而且可以直接被面向对象的程序设计语言所接受。

总之，面向对象方法是以认识论为基础，用对象来理解和分析问题空间，并设计和开发由对象构成的软件系统（解空间）的方法。由于问题空间和解空间都是由对象组成的，这样可以消除由于问题空间和求解空间结构上的不一致带来的问题。简言之，面向对象就是面向事情本身，面向对象的分析过程就是认识客观世界的过程。

1.3.3 面向对象方法的开发过程

面向对象的软件开发方法一般包括面向对象分析、面向对象设计和面向对象编程三个环节。

一般将用面向对象方法开发软件的工作过程分为四个阶段。

（1）系统调查和需求分析。对系统将要面临的具体管理问题以及用户对系统开发的需求进行调查研究，即先弄清要干什么的问题。

（2）分析问题的性质和求解问题。在繁杂的问题中抽象地识别出对象以及其行为、结构、属性、方法等。这一阶段一般称为面向对象分析，简称为 OOA。

（3）整理问题。即对分析的结果作进一步地抽象、归类、整理，并最终以范式的形式将它们确定下来。这一阶段一般称为面向对象设计，简称为 OOD。

（4）程序实现。即用面向对象的程序设计语言将上一步整理的范式直接映射（即直接用程序语言来取代）为应用程序软件。这一阶段一般称为面向对象的程序设计，简称为 OOP。

1. 面向对象的分析

面向对象的分析是面向对象方法的一个组成部分。在一个系统的开发过程中进行了系统业务调查以后，就可以按照面向对象的思想来分析问题。应当注意的是，面向对象的分析方法所说的分析与结构化分析有较大的区别。面向对象的分析方法所强调的是在系统调查资料的基础上，针对面向对象方法所需要的素材进行的归类分析和整理，而不是对管理业务现状和方法的分析，这点也正是面向对象的方法在信息系统开发过程中的关键点。

1）面向对象分析的原则

用面向对象的方法对所调查结果进行分析处理时，一般依据以下几项原则。

(1) 抽象(abstraction)。指为了某一分析目的而集中精力研究对象的某一性质,可以忽略其他与此目的无关的部分。

抽象机制被用在数据分析方面,称为数据抽象。数据抽象是面向对象分析的核心。数据抽象把一组数据对象以及作用其上的操作组成一个程序实体。使得外部只知道它是如何表示的。在应用数据抽象原理时,系统分析人员必须确定对象的属性以及处理这些属性的方法,并借助于方法获得属性,在面向对象的分析中属性和方法被认为是不可分割的整体。

抽象机制也被用在对过程的分解方面,称为过程抽象。恰当的过程抽象可以对复杂过程的分解和确定以及描述对象发挥积极的作用。

(2) 封装(encapsulation)。即信息隐蔽,是指在确定系统的某一部分内容时,还要考虑到其他部分的信息和联系也都在这一部分的内部进行,外界各部分之间的信息联系应尽可能地少。

(3) 继承(inheritance)。指能直接获得已有的性质和特征而不必重复定义它们。面向对象的分析可以一次性地指定对象的公共属性和方法,然后再特殊化或扩展这些属性和方法使之转化为特殊情况,这样可大大地减轻在系统实现过程中的重复劳动。在共有属性的基础之上,继承者也可以定义自己独有的特性。

(4) 相关(association)。意指联合(union)或连接(connection),指把某一时刻或相同环境下发生的事物联系在一起。

(5) 消息通信(communication with message)。指在对象之间互相传递信息的通信方式。

(6) 组织方法(method of organization)。在分析和认识世界时,可综合用如下三种组织方法:

① 特定对象与其属性之间的区别;
② 整体对象与相应组成部分对象之间的区别;
③ 不同对象类的构成及其区别。

(7) 比例(scale)。一种运用整体与部分原则,辅助处理复杂问题的方法。

(8) 行为范畴(categories of behavior)。很多情况下是针对被分析对象而言的,它们主要包括:

① 基于直接原因的行为。
② 时变性行为。
③ 功能查询性行为。

2) 面向对象的分析过程

图 1.11 面向对象的分析抽象的方法

面向对象的分析方法是建立在对处理对象客观运行状态的信息模拟(实体关系图和语义数据模型)和面向对象程序设计语言的概念基础之上,这种关系可以形象地用图 1.11 表示。

面向对象的分析方法从信息模拟中借鉴了属性、关系、结构和对象等概念,这些概念在问题域中是用来描述某些事物和实体的；从面向对象的程序设计语言中吸取了封装、分类结构和继承性等概念。面向对象的分析强调以下三点：

（1）在分析和规格说明的总体框架中始终贯穿着结构化方法。

（2）用户和系统之间、系统中实体之间的相互通信用消息实现。

（3）在总体框架中对每个部分提供的方法和性能进行分类。

利用面向对象的分析方法对一个具体事物进行分析时，一般可分为以下五个步骤：

（1）确定主题。这里所说的主题是指事物的总体概貌和总体分析模型。

（2）确定对象和类。对象是对数据及其处理方式的抽象，它反映的是系统保存和处理现实世界中某些事物的信息的能力。类是多个对象的共同属性和方法集合的描述，它包括如何在一个类中建立一个新对象的描述。

（3）确定结构。结构是指问题域的复杂性和连接关系。类成员结构反映了泛化-特化关系，整体-部分结构反映整体和局部之间的关系。

（4）确定属性。这里所说的属性就是数据元素，可用来描述对象或分类结构的实例，并在对象的存储中指定。

（5）确定方法。这里所说的方法是在收到消息后必须进行的一些处理方法，方法要在图中定义，并在对象的存储中指定。对于每个对象和结构来说，那些用来增加、修改、删除和选择一个方法本身都是隐含的，而有些则是显示的，如计算费用等。

3）面向对象的分析模型

面向对象的分析就是把问题空间分解成一些类或对象，找出这些对象的特点（即属性和服务），以及对象间的关系（一般特殊、整体部分关系），并由此产生一个规格说明。所面临的挑战是问题空间的理解、人与人之间的通信和需求的不断变化。

2．面向对象设计

面向对象的设计是面向对象方法中一个中间过渡环节。它是把分析阶段得到的对目标系统的需求转变成符合成本和质量要求的、抽象的系统实现方案的过程。其主要作用是对面向对象分析的结果作进一步的规范化整理，以便能够被面向对象程序设计直接接受。

从面向对象分析到面向对象设计，是一个逐渐扩充模型的过程，即面向对象设计就是利用面向对象观点建立求解域模型的过程，是一个多次反复迭代的过程。

1）面向对象设计的准则

在面向对象设计过程中应遵循如下准则。

（1）模块化：对象是面向对象软件系统中的模块。它是把数据结构和操作这些数据的方法紧密地结合在一起所构成的模块。

（2）抽象：面向对象的方法不仅支持过程抽象，还支持数据抽象。对象类实际上是具有继承机制的抽象数据类型。

（3）信息隐藏：信息隐藏通过对象的封装来实现。类结构分离了接口与实现，从而支持了信息隐藏。

（4）弱耦合：对象之间的耦合通过消息连接来实现，尽量降低消息连接的复杂程度。

（5）强内聚：一个服务应该完成一个且仅完成一个功能；一个类应该只有一个用途，它的属性和服务应该是高内聚的。

(6) 可重用：重用是提高软件开发生产率和目标系统质量的重要途径。在设计时应尽量使用已有的类，如果确实需要创建新类，应该考虑将来的可重复使用性。

2) 面向对象设计的组成

面向对象设计可分为以下四部分：

(1) 问题空间部分的设计(PDC)。

通过面向对象分析所得出的问题域精确模型，为设计问题域子系统奠定了良好的基础，建立了完整的框架。通常，面向对象设计仅需对问题域模型作一些补充或修改，主要是增添、合并或分解类、属性及服务，调整继承关系等。

当问题域子系统过于复杂庞大时，应该把它进一步分解成若干个更小的子系统。

问题域子系统设计也称为主题部件设计。

(2) 人机交互部分的设计(HIC)。

在面向对象分析阶段已经对用户的界面需求做了初步分析，在面向对象设计过程中，应该对目标系统的人机交互子系统进行相应设计，以确定人机交互界面的细节，其中包括指定窗口和报表的形式、设计命令层次等内容。根据需求把交互细节加入用户界面设计中，包括人机交互所必需的实际显示和输入。

在该设计过程中使用原型支持的系统化的设计策略，是成功地设计人机交互子系统的关键。

(3) 任务管理部分的设计(TMC)。

任务管理子系统就是将重要的服务设计成任务类。

任务管理部分的设计包括确定各类任务，把任务分配给适当的硬件或软件去执行。

常见的任务有事件驱动型任务、时钟驱动型任务、优先任务、关键任务和协调任务等类型。

(4) 数据管理部分的设计(DMC)。

数据管理部分提供了在数据管理系统中存储和检索对象的基本结构，包括对永久性数据的访问和管理。它分离了数据管理机构所关心的事项，包括文件、关系型 DBMS 或面向对象 DBMS 等。

不同的数据存储管理模式有不同的特点，适用范围也不相同，设计者应该根据应用系统的特点选择适用的模式。数据管理方法包括文件管理、关系数据库管理和面向对象库数据管理。

在以上四部分中，每个部分划分成五个层次：主题层、类和对象层、结构层、属性层和服务层。即每个部分的设计分为以上五个阶段。它们与分析阶段是对应的。对应关系如表 1.3 所示。

表 1.3 分析阶段和设计阶段的对应

序号	分析阶段	设计阶段
1	定义主题	主题层
2	标识对象	类和对象层
3	标识对象所属类	结构层
4	标识对象的属性	属性层
5	标识对象的行为	服务层

1.3.4 面向对象的建模语言——统一建模语言

1. 统一建模语言

统一建模语言(Unified Modeling Language,UML)是一种支持对象技术的建模语言,也是一个通用的标准建模语言,可以对任何具有静态结构和动态行为的系统进行建模。在这里统一表示是一种通用的标准,称为软件工业界的一种标准,UML表述的内容能被各类人员所理解,包括客户、领域专家、分析师、设计师、程序员、测试工程师及培训人员等;建模代表建立软件系统的模型;语言表明它是一套按照特定规则和模式组成的符号系统,它用半形式化方法定义,即用图形符号、自然语言和形式语言相结合的方法来描述定义。

UML展现了一系列最佳工程实践,在对大规模、复杂系统进行建模方面,特别是在软件架构层次是十分有效。UML最适于数据建模、业务建模、对象建模和组件建模。UML作为一种模型语言,它使开发人员专注于建立产品的模型和结构,而不是选用什么程序语言和算法实现。实现UML建模需要CASE工具,这类工具主要有Rational Rose、Together、Visual Modeler等。

2. 统一建模语言的描述

利用统一建模语言进行建模就是用统一建模语言中的各种语言符号按一定的规则进行相关描述或进行表述。了解统一建模语言中的各种语言符号是十分必要的。统一建模语言中的各种语言符号及说明如表1.4所示。

表1.4 统一建模语言中的各种语言符号及说明

序号	图素名称	表示含义	代表符号
1	类	对一组具有相同属性、相同操作、相同关系和相同语义的对象的描述	NewClass
2	对象		对象名:类名 :类名 类名
3	接口	描述一个类或构件的一个服务的操作集	Interface
4	协作	定义了一个交互,它是由一组共同工作以提供某种协作行为的角色和其他元素构成的一个群体	
5	用例	对一组动作序列的描述	
6	主动类	对象至少拥有一个进程或线程的类	class $suspend() $flush()
7	构件	系统中物理的、可替代的部件	componet
8	参与者	在系统外部与系统直接交互的人或事物	actor
9	节点	在运行时存在的物理元素	NewProcessor

续表

序号	图素名称	表 示 含 义	代表符号
10	交互	由在特定语境中共同完成一定任务的一组对象间交换的消息组成	→
11	状态机	描述了一个对象或一个交互在生命期内响应事件所经历的状态序列	state
12	包	把元素组织成组的机制	NewPackage
13	注释事物	UML 模型的解释部分	
14	依赖	一条可能有方向的虚线	---→
15	关联	一条实线,可能有方向	
16	泛化	一条带有空心箭头的实线	——▷
17	实现	一条带有空心箭头的虚线	---▷

统一建模语言的重要内容可以由下列五类图共 10 种图来定义。

1) 用例图

用例图从用户的角度描述系统功能。它将系统划分为对用户有意义的事务,这些事务称为用例(use case),用户称为角色(actor)(见图 1.12)。

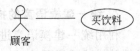

图 1.12 用例图示例

2) 静态图

静态图展现系统的静态结构组成及特征,包括类图、对象图和包图,如图 1.13～图 1.15 所示。

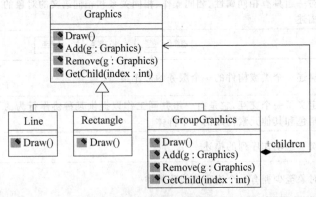

图 1.13 类示意图

类图描述系统中类的静态结构,在其定义中不仅包括系统中的类,表示类中的联系,如关联、依赖、聚合等,也包括类的内部结构(类的属性和操作)。

对象图是类图的示例,对象图只能在系统的某一时间段存在。

包图由包或类组成,表示包与包之间的关系,包图用来描述系统的分层结构。

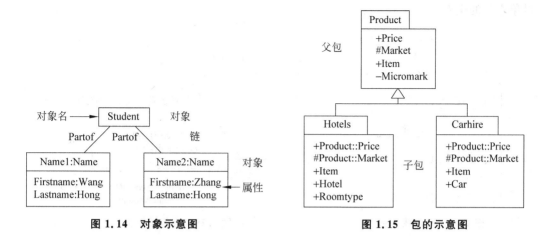

图 1.14　对象示意图　　　　图 1.15　包的示意图

3）行为图

行为图描述系统的动态模型和组成对象间的交互关系,包括状态图和活动图,如图 1.16 和图 1.17 所示。

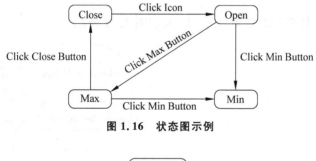

图 1.16　状态图示例

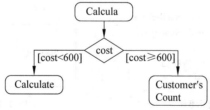

图 1.17　活动图示例

状态图描述类的对象所有可能的状态以及事件发生时状态的转移条件。状态图是对类图的补充,可为那些有多个状态、其行为受外界环境的影响并且发生改变的类画状态图。

活动图描述满足用例要求所要进行的活动以及活动间的约束关系。活动图的状态代表了运算执行的状态。

4）交互图

交互图描述对象间交互关系,包括顺序图和协作图,如图 1.18 和图 1.19 所示。

顺序图显示对象间的动态协作合作关系,它强调对象之间消息发送的顺序,同时显示对象之间的交互。

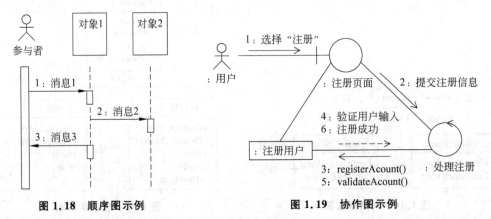

图 1.18 顺序图示例　　　　图 1.19 协作图示例

协作图描述对象间的协作关系，协作图与顺序图相似，显示对象间的动态协作关系，还显示对象以及它们之间的关系。

如果强调时间和顺序，则用顺序图；如果强调通信关系，则偏向用协作图。

5) 实现图

实现图包括构件图和部署图，如图 1.20 和图 1.21 所示。

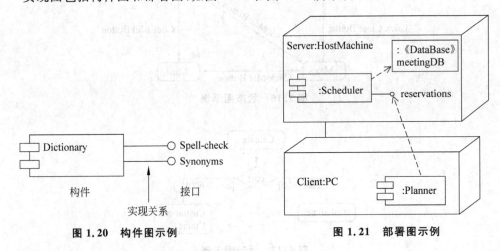

图 1.20 构件图示例　　　　图 1.21 部署图示例

构件图描述代码构件的物理结构和各构件之间的依赖关系。构件图有利于分析和理解构件之间的相互影响程度。

部署图定义系统中软件和硬件的物理体系结构，它显示实际的计算机和设备以及它们之间的连接关系，也可以表示连接的设备类型和构件之间的依赖性。在节点内部，放置可执行部件和对象以显示节点跟可执行软件单元的对应关系。

图 1.22 为各图之间的关系和使用阶段。

3. 基于 UML 的对象的分析与设计

面向对象开发过程通常是一个迭代的过程。面向对象开发过程中的分析阶段和设计阶段没有明显的界限。因为两个阶段都是围绕对象模型展开，在分析阶段所得到的模型称为分析类图;在设计阶段所得到的模型称为设计类图。分析类图的获得是通过构造静态模型而得到;设计类图的获得是在分析类图的基础上通过构造动态模型而获得;因此,在面向对象开发中,静态模型和动态模型都是为了构造完整对象模型而服务的。

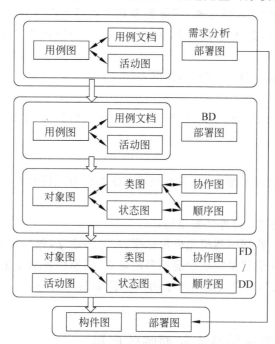

图 1.22 各图之间的关系和使用阶段

以 UML 的各种图的使用来说明面向对象的分析与设计过程如下。

1) 需求调查分析

需求调查分析的结果一般用文字描述,必要时也可用业务流程图和部署图辅助描述。

2) 用例建模

用例建模主要是完成用例模型的建立。主要任务有:

(1) 确定执行者。通过对系统需求陈述的分析,可以确定系统的执行者。

(2) 确定用例。在确定执行者之后,结合相关的领域知识,进一步分析系统的需求,可以确定系统的用例,用例的简要描述。

(3) 确定用例之间的关系。确定执行者和用例之后,进一步确定用例之间的关系。

这部分的结果可用用例图、场景描述和活动图等表述。

3) 静态建模

静态建模完成静态模型的建立。首先给出候选的对象类。然后,经过标识责任、标识协作者和复审,定义类的属性、操作和类之间的关系。

这部分的结果主要用对象图、类图和包图等来表述。

4) 系统设计

系统设计完成系统各组成部分以及组成部分之间的关系的设计,确定实现系统的策略和目标系统的高层结构。用包图等描述系统设计。

5) 对象设计

对象设计主要完成对象模型的建立。对象设计主要有两个任务：

(1) 对类的属性和操作的实现细节进行设计（分析阶段）。

(2) 分别从人机交互、数据管理、任务管理和问题域方面考虑，以实现的角度添加一些类或优化类的结构，包括确定解空间中的类、关联、接口形式及实现服务的算法（设计阶段）。

这部分结果可用对象图、类图和构件图表述。

6) 动态建模

动态建模是建立动态模型，动态模型明确规定什么时候做，主要由状态图、活动图、顺序图和协作图描述。

7) 物理建模

物理建模是建立物理模型，物理模型表述系统物理节点的分布，主要用部署图和构件图描述。

为了更好地开发系统，另外还要进行：

(1) 数据库建模。数据库建模完成数据库、表等的建立，包括相关的优化。

(2) 功能建模。功能建模完成功能模型的建立，该模型表示变化的系统"功能"性质，它指明系统应"做什么"，更直接地反映了用户对目标系统的需求。功能模型由一组数据流图组成。

1.3.5 面向对象方法的特点和面临的问题

OO方法以对象为基础，利用特定的软件工具直接完成从对象客体的描述到软件结构之间的转换。这是OO方法最主要的特点和成就。OO方法的应用解决了传统结构化开发方法中客观世界描述工具与软件结构的不一致性问题，缩短了开发周期，解决了从分析和设计到软件模块结构之间多次转换映射的繁杂过程，是一种很有发展前途的系统开发方法。但是同原型方法一样，OO方法需要一定的软件基础支持才可以应用，另外在大型的MIS开发中如果不经自顶向下的整体划分，而是一开始就自底向上地采用OO方法开发系统，同样也会造成系统结构不合理、各部分关系失调等问题。所以笔者认为OO方法和结构化方法目前仍是两种在系统开发领域相互依存、不可替代的方法。

第 2 章

开发技术和环境简介

2.1 SQL Server 2008 简介

2.1.1 SQL 简介

SQL(Structured Query Language)被翻译为结构化查询语言,是一种数据库查询和程序设计语言,用于存取数据以及查询、更新和管理关系数据库系统;同时也是数据库脚本文件的扩展名。

结构化查询语言是高级的非过程化编程语言,允许用户在高层数据结构上工作。

结构化查询语言最早是 IBM 的圣约瑟研究实验室为其关系数据库管理系统 SYSTEM R 开发的一种查询语言,它的前身是 SQUARE 语言。

SQL 语言结构简洁,功能强大,简单易学,自从 IBM 公司 1981 年推出以来,SQL 语言得到了广泛的应用。如今无论是像 Oracle、Sybase、DB2、Informix、SQL Server 这些大型的数据库管理系统,还是像 Visual FoxPro、PowerBuilder 这些 PC 上常用的微小型数据库系统,都支持 SQL 语言作为查询语言。

SQL 主要有以下特点。

(1) 一体化:该语言集数据定义(DDL)、数据操纵(DML)和数据控制(DCL)于一体,可以完成数据库中的全部工作。

(2) 使用方式灵活:它具有两种使用方式,既可以直接以命令方式交互使用,也可以嵌入如 C、C++、C#、VB、FORTRAN、COBOL、Java 等宿主语言中使用。

(3) 非过程化:使用时只需要告诉计算机"做什么",不需要告诉计算机"如何做"。

(4) 语言简洁,语法简单,好学好用:在 ANSI 标准中,只包含了 94 个英文单词,核心功能只用 6 个动词,语法接近英语口语。

结构化查询语言包含以下六部分。

(1) 数据查询语言(DQL):其语句动词为 SELECT,用来从表中获得数据,确定数据怎样在应用程序给出。保留字 SELECT 是 DQL 和 SQL 用得最多的动词,与此搭配使用的其他 DQL 常用的保留字有 WHERE、ORDER BY、GROUP BY 和 HAVING。由这些保留字构成复杂的 DQL 语句,再与其他类型的 SQL 语句一起使用构成更复杂的 SQL 语句或程序。

(2) 数据操作语言(DML)：其语句包括动词 INSERT、UPDATE 和 DELETE，它们分别用于添加、修改和删除表中的行。

(3) 事务处理语言(TPL)：其语句能确保被 DML 语句影响的表的所有行及时得以更新或恢复。TPL 语句包括 BEGIN TRANSACTION、COMMIT 和 ROLLBACK，它们分别用于开始一个事务，提交一个事务和回滚一事务。

(4) 数据控制语言(DCL)：其语句主要有 GRANT 和 REVOKE，用以授予或回收用户和用户组对数据库对象的访问权限。

(5) 数据定义语言(DDL)：其语句主要包括动词 CREATE、ALTER 和 DROP。在数据库中创建新、修改或删除对象(如数据库、表、视图、索引等)。

(6) 指针控制语言(CCL)：它的语句，像 DECLARE CURSOR、FETCH INTO 和 UPDATE WHERE CURRENT 用于对一个或多个表单进行操作。

2.1.2 SQL Server 的发展

SQL Server 是微软公司推出的 SQL 的系列产品的名称，从 1988 年微软推出第一个 SQL Server 版本到今天的 SQL Server 2012 版本，经历 20 多年发展过程，推出了约 10 个版本的产品，表 2.1 概述了这一发展历程。

表 2.1 SQL Server 发展历程

年份	版本	说明
1988	SQL Server	与 Sybase 共同开发的、运行于 OS/2 上的联合应用程序
1993	SQL Server 4.2 一种桌面数据库	一种功能较少的桌面数据库，能够满足小部门数据存储和处理的需求。数据库与 Windows 集成，界面易于使用并广受欢迎
1994		微软与 Sybase 终止合作关系
1995	SQL Server 6.05 一种小型商业数据库	对核心数据库引擎做了重大的改写。这是首次"意义非凡"的发布，性能得以提升，重要的特性得到增强。在性能和特性上，尽管以后的版本还有很长的路要走，但这一版本的 SQL Server 具备了处理小型电子商务和内联网应用程序的能力，而在花费上却少于其他的同类产品
1996	SQL Server 6.5	SQL Server 逐渐突显实力，以至于 Oracle 推出了运行于 NT 平台上的 7.1 版本作为直接的竞争
1998	SQL Server 7.0 一种 Web 数据库	再一次对核心数据库引擎进行了重大改写。这是相当强大的、具有丰富特性的数据库产品的明确发布，该数据库介于基本的桌面数据库(如 Microsoft Access)与高端企业级数据库(如 Oracle 和 DB2)之间(价格上亦如此)，为中小型企业提供了切实可行(并且还廉价)的可选方案。该版本易于使用，并提供了对于其他竞争数据库来说需要额外附加的昂贵的重要商业工具(例如，分析服务、数据转换服务)，因此获得了良好的声誉

续表

年份	版本	说明
2000	SQL Server 2000 一种企业级数据库	SQL Server 在可扩缩性和可靠性上有了很大的改进,成为企业级数据库市场中重要的一员(支持企业的联机操作,其所支持的企业有 NASDAQ、戴尔和巴诺等)。虽然 SQL Server 在价格上有很大的上涨(尽管算起来还只是 Oracle 售价的一半左右),减缓了其最初被接纳的进度,但它卓越的管理工具、开发工具和分析工具赢得了新的客户。2001 年,在 Windows 数据库市场(2001 年价值 25.5 亿美元),Oracle(34%的市场份额)不敌 SQL Server(40%的市场份额),最终将其市场第一的位置让出。2002 年,差距继续拉大,SQL Server 取得 45%的市场份额,而 Oracle 的市场份额下滑至 27%(来源于 2003 年 5 月 21 日的 Gartner Report)
2005	SQL Server 2005	对 SQL Server 的许多地方进行了改写,例如,通过名为集成服务(Integration Service)的工具来加载数据,不过,SQL Server 2005 最伟大的飞跃是引入了.NET Framework。引入.NET Framework 将允许构建.NET SQL Server 专有对象,从而使 SQL Server 具有灵活的功能,正如包含 Java 的 Oracle 所拥有的那样
2008	SQL Server 2008	SQL Server 2008 以处理目前能够采用的许多种不同的数据形式为目的,通过提供新的数据类型和使用语言集成查询(LINQ),在 SQL Server 2005 的架构的基础之上打造出了 SQL Server 2008。SQL Server 2008 同样涉及处理像 XML 这样的数据、紧凑设备(compact device)以及位于多个不同地方的数据库安装。另外,它提供了在一个框架中设置规则的能力,以确保数据库和对象符合定义的标准,并且,当这些对象不符合该标准时,还能够就此进行报告
2012	SQL Server 2012	作为云就绪信息平台中的关键组件,可以帮助企业释放突破性的业务洞察力;它对关键业务充满信心,能够快速地构建相应的解决方案来实现本地和公有云之间的数据扩展

2.1.3 SQL Server 2008 的版本

微软公司推出了五种不同的 SQL Server 版本,各本版说明如表 2.2 所示。

表 2.2 SQL Server 2008 的不同版本说明表

版本	说明	适用范围
企业版 Enterprise Edition	一个全面的数据管理和商业智能平台,提供企业级的可扩展性、高可用性和高安全性以运行企业关键业务应用	大规模联机事务处理(OLTP) 大规模报表 先进的分析 数据仓库
标准版 Standard Edition	一个完整的数据管理和商业智能平台,提供最好的易用性和可管理性来运行部门级应用	部门级应用 中小型规模 OLTP 报表和分析
工作组版 Workgroup Edition	一个可信赖的数据管理和报表平台,提供各分支应用程序以安全、远程同步和管理功能	分支数据存储 分支报表 远程同步

续表

版 本	说 明	适 用 范 围
学习版 Express Edition	提供学习和创建桌面应用程序和小型应用程序，并可被 ISVs 重新发布的免费版本	入门级 & 学习 免费的 ISVs 重发 丰富的桌面应用
移动版 Compact Edition	一个免费的嵌入式 SQL Server 数据库，可创建移动设备、桌面端和 Web 端独立运行的和偶尔连接的应用程序	独立嵌入式开发 断开式连接客户端

2.1.4 SQL Server 2008 的新增功能

SQL Server 2008 在 Microsoft 的数据平台上发布，它可以将结构化、半结构化和非结构化文档的数据直接存储到数据库中。

SQL Server 2008 提供一系列丰富的集成服务，可以对数据进行查询、搜索、同步、报告和分析之类的操作。数据可以存储在各种设备上，从数据中心最大的服务器一直到桌面计算机和移动设备，用户都可以控制数据而不用管数据存储在哪里。

SQL Server 2008 允许用户在使用 Microsoft .NET 和 Visual Studio 开发的自定义应用程序中使用数据，在面向服务的架构（SOA）和通过 Microsoft BizTalk Server 进行的业务流程中使用数据。用户可以通过他们日常使用的工具（例如 2007 Microsoft Office 系统）直接访问数据。SQL Server 2008 提供一个可信的、高效率智能数据平台，可以满足用户的所有数据需求。

SQL Server 2008 所提供的新的主要功能包括如下几方面。

1. 保护有价值的信息

1) 透明的数据加密

允许加密整个数据库、数据文件或日志文件，无须更改应用程序。这样做的好处包括：同时使用范围和模糊搜索来搜索加密的数据，从未经授权的用户搜索安全的数据，可以不更改现有应用程序的情况下进行数据加密。

2) 可扩展的键管理

SQL Server 2005 为加密和键管理提供一个全面的解决方案。SQL Server 2008 通过支持第三方键管理和 HSM 产品提供一个优秀的解决方案，以满足不断增长的需求。

3) 审计

通过 DDL 创建和管理审计，同时通过提供更全面的数据审计来简化遵从性。这允许组织回答常见的问题，例如"检索什么数据？"。

2. 确保业务连续性

1) 增强的数据库镜像

SQL Server 2008 构建于 SQL Server 2005 之上，但增强的数据库镜像，包括自动页修复、提高性能和提高支持能力，因而是一个更加可靠的平台。

2）数据页的自动恢复

SQL Server 2008 允许主机器和镜像机器从 823/824 类型的数据页错误透明地恢复，它可以从透明于终端用户和应用程序的镜像伙伴请求新副本。

3）日志流压缩

数据库镜像需要在镜像实现的参与方之间进行数据传输。使用 SQL Server 2008，参与方之间的输出日志流压缩提供最佳性能，并最小化数据库镜像使用的网络带宽。

3．启用可预测的响应

1）资源管理者

通过引入资源管理者来提供一致且可预测的响应，允许组织为不同的工作负荷定义资源限制和优先级，这允许并发工作负荷为它们的终端用户提供一致的性能。

2）可预测的查询性能

通过提供功能锁定查询计划支持更高的查询性能稳定性和可预测性，允许组织在硬件服务器替换、服务器升级和生产部署之间推进稳定的查询计划。

3）数据压缩

更有效地存储数据，并减少数据的存储需求。

4）热添加 CPU

允许 CPU 资源在支持的硬件平台上添加到 SQL Server 2008，以动态调节数据库大小而不强制应用程序宕机。注意，SQL Server 已经支持在线添加内存资源的能力。

为了抓住如今风云变幻的商业机会，公司需要具备快速创建和部署数据驱动的解决方案的能力。SQL Server 2008 减少了管理和开发应用程序的时间和成本。

4．按照策略进行管理

1）Policy-Based Management

Policy-Based Management 是一个基于策略的系统，用于管理 SQL Server 2008 的一个或多个实例。将其与 SQL Server Management Studio 一起使用可以创建管理服务器实体（比如 SQL Server 实例、数据库和其他 SQL Server 对象）的策略。

2）精简的安装

SQL Server 2008 通过重新设计安装、设置和配置体系结构，对 SQL Server 服务生命周期进行了巨大的改进。这些改进将物理定位在硬件上的安装与 SQL Server 软件的配置隔离，允许组织和软件合作伙伴提供推荐的安装配置。

3）性能数据收集

性能调节和故障诊断对于管理员来说是一项耗时的任务。为了给管理员提供可操作的性能检查，SQL Server 2008 包含更多详尽性能数据的集合，一个用于存储性能数据的集中化的新数据仓库，以及用于报告和监视的新工具。

5．简化应用程序开发

1）语言集成查询（LINQ）

开发人员可以使用诸如 C♯ 或 VB.NET 等托管的编程语言而不是 SQL 语句查询数据。允许根据 ADO.NET（LINQ to SQL）、ADO.NET DataSets（LINQ to DataSet）、

ADO.NET Entity Framework(LINQ to Entities)，以及实体数据服务映射供应商运行 .NET语言编写的无缝、强类型、面向集合的查询。新的 LINQ to SQL 供应商允许开发人员在 SQL Server 2008 表和列上直接使用 LINQ。

2）ADO.NET Object Services

ADO.NET 的 Object Services 层将具体化、更改跟踪和数据持久作为 CLR 对象。使用 ADO.NET 框架的开发人员可以使用 ADO.NET 管理的 CLR 对象进行数据库编程。SQL Server 2008 引入更有效、优化的支持来提高性能和简化开发。

6．存储任何信息

1）DATE/TIME

SQL Server 2008 引入新的日期和时间数据类型：

- DATE——仅表示日期的类型。
- TIME——仅表示时间的类型。
- DATETIMEOFFSET——可以感知时区的 datetime 类型。
- DATETIME2——比现有 DATETIME 类型具有更大小数位和年份范围的 datetime 类型。

新的数据类型允许应用程序拥有独立的日期和时间类型，同时为时间值提供大的数据范围或用户定义的精度。

2）HIERARCHY ID

允许数据库应用程序使用比当前更有效的方法制定树结构的模型。新的系统类型 HierarchyId 可以存储代表层次结构树中节点的值。这种新类型将作为一种 CLR UDT 实现，将暴露几种有效并有用的内置方法，用于使用灵活的编程模型创建和操作层次结构节点。

3）FILESTREAM Data

允许大型二进制数据直接存储在 NTFS 文件系统中，同时保留数据库的主要部分并维持事务一致性。允许扩充传统上由数据库管理的大型二进制数据，可以存储到数据库外部更经济实惠的存储设备上，且没有泄密风险。

4）集成的全文本搜索

集成的全文本搜索使文本搜索和关系型数据之间能够无缝转换，同时允许用户使用文本索引在大型文本列上执行高速文本搜索。

5）Sparse Columns

NULL 数据不占据物理空间，提供高效的方法来管理数据库中的空数据。例如，Sparse Columns 允许通常有许多空值的对象模型存储在 SQL Server 2005 数据库中，而无须耗费大量空间成本。

6）大型用户定义的类型

SQL Server 2008 消除用户定义类型（UDT）的 8 KB 限制，允许用户极大地扩展其 UDT 的大小。

7）空间数据类型

通过使用对空间数据的支持，将空间能力构建到应用程序中。

使用地理数据类型实现"圆面地球"解决方案,使用经纬度来定义地球表面的区域。

使用地理数据类型实现"平面地球"解决方案,存储与投影平面表面和自然平面数据关联的多边形、点和线,例如内部空间。

SQL Server 2008 提供全面的平台,在用户需要的时候提供智能。

7. 集成任何数据

1) 备份压缩

在线保存基于磁盘的备份昂贵且耗时。借助 SQL Server 2008 备份压缩,在线保存备份所需的存储空间更少,备份运行速度更快,因为需要的磁盘 I/O 更少。

2) 已分区表并行

分区允许组织更有效地管理增长迅速的表,可以将这些表透明地分成易于管理的数据块。SQL Server 2008 继承了 SQL Server 2005 中的分区优势,但提高了大型分区表的性能。

3) 星状连接查询优化

SQL Server 2008 为常见的数据仓库场景提供改进的查询性能。星状连接查询优化通过识别数据仓库连接模式来减少查询响应时间。

4) Grouping Sets

Grouping Sets 是对 GROUP BY 子句的扩展,允许用户在同一个查询中定义多个分组。Grouping Sets 生成单个结果集(等价于不同分组行的一个 UNION ALL),使得聚集查询和报告变得更加简单快速。

5) 更改数据捕获

使用"更改数据捕获",可以捕获更改内容并存放在更改表中。它捕获完整的更改内容,维护表的一致性,甚至还能捕获跨模式的更改。这使组织可以将最新的信息集成到数据仓库中。

6) MERGE SQL 语句

随着 MERGE SQL 语句的引入,开发人员可以更加高效地处理常见的数据仓库存储应用场景,比如检查某行是否存在,然后执行插入或更新。

7) SQL Server Integration Services(SSIS)管道线改进

"数据集成"包现在可以更有效地扩展,可以利用可用资源和管理最大的企业规模工作负载。新的设计将运行时的伸缩能力提高到多个处理器。

8) SQL Server Integration Services(SSIS)持久查找

执行查找的需求是最常见的 ETL 操作之一。这在数据仓库中特别普遍,其中事实记录需要使用查找将企业关键字转换成相应的替代字。SSIS 增强查找的性能以支持最大的表。

8. 发布相关的信息

1) 分析规模和性能

SQL Server 2008 使用增强的分析能力和更复杂的计算和聚集交付更广泛的分析。新的立方体设计工具帮助用户精简分析基础设施的开发,让他们能够为优化的性能构建解决方案。

2) 块计算

块计算在处理性能方面提供极大的改进,允许用户增加其层次结构的深度和计算的复杂性。

3) 写回

新的 MOLAP 在 SQL Server 2008 Analysis Services 中启用写回(writeback)功能,不再需要查询 ROLAP 分区。这为用户提供分析应用程序中增强的写回场景,而不牺牲传统的 OLAP 性能。

9. 推动可操作的商务洞察力

1) 企业报表引擎

报表可以使用简化的部署和配置在组织中方便地分发(内部和外部),这使得用户可以方便地创建和共享任何规格和复杂度的报表。

2) Internet 报表部署

通过在 Internet 上部署报表,很容易找到客户和供应商。

3) 管理报表体系结构

通过集中化存储和所有配置设置的 API,使用内存管理、基础设施巩固和更简单的配置来增强支持能力和控制服务器行为的能力。

4) Report Builder 增强

通过报表设计器轻松构建任何结构的特殊报表和创作报表。

5) 内置的表单认证

内置的表单认证让用户可以在 Windows 和 Forms 之间方便地切换。

6) 报表服务器应用程序嵌入

报表服务器应用程序嵌入使得报表和订阅中的 URL 可以重新指向前端应用程序。

7) Microsoft Office 集成

SQL Server 2008 提供新的 Word 渲染,允许用户通过 Microsoft Office Word 直接使用报表。此外,现有的 Excel 渲染器已经得以极大的增强,以支持嵌套的数据区域、子报表以及合并的表格改进等功能。这让用户保持布局保真度并改进 Microsoft Office 应用程序对报表的总体使用。

8) 预测性分析

SQL Server Analysis Services 继续交付高级的数据挖掘技术。更好的时间序列支持增强了预测能力。增强的挖掘结构提供更大的灵活性,可以通过过滤执行集中分析,还可以提供超出挖掘模型范围的完整信息报表。新的交叉验证允许同时确认可信结果的精确性和稳定性。此外,针对 Office 2007 的 SQL Server 2008 数据挖掘附件提供的新特性使组织中的每个用户都可以在桌面上获得更多可操作的洞察。

2.1.5 SQL Server 2008 的新增特点

SQL Server 2008 具有以下特点。

(1) 可信任的——使公司可以很高的安全性、可靠性和可扩展性来运行它们最关键任务的应用程序。

SQL Server 2008 为关键任务应用程序提供了强大的安全特性、可靠性和可扩展性。

① 保护你的信息。

② 确保业务可持续性。

③ 最佳的和可预测的系统性能。

(2) 高效的——使公司可以降低开发和管理它们的数据基础设施的时间和成本。

SQL Server 2008 降低了管理系统、.NET 架构和 Visual Studio Team System 的时间和成本,使开发人员可以开发强大的下一代数据库应用程序。

① 基于政策的管理。

② 改进了安装。

③ 加速开发过程。

④ 偶尔连接系统。

⑤ 不只是关系数据。

(3) 智能的——提供了一个全面的平台,可以在用户需要的时候给他发送观察和信息。

SQL Server 2008 提供了一个全面的平台,用于当用户需要时可以为其提供智能化。

① 集成任何数据。

② 发送相应的报表。

③ 使用户获得全面的洞察力。

2.1.6 SQL Server 2008 安装要求

SQL Server 2008 对硬件、软件和网络环境都有相应要求。具体要求见表 2.3。

表 2.3 SQL Server 2008 对硬件、软件和网络环境都有相应要求

名 称	安 装 要 求
CPU	建议的最低要求是 32 位版本对应 1GHz 的处理器,64 位版本对应 1.6GHz 的处理器,或兼容的处理器,或具有类似处理能力的处理器,但推荐使用 2GHz 的处理器
硬盘	数据库引擎和数据文件、复制以及全文搜索 280MB;Analysis Services 和数据文件 90MB;Reporting Services 和报表管理器 120MB;Integration Services 120MB;客户端组件 850MB;SQL Server 联机丛书和 SQL Server Compact 联机丛书 240MB
驱动器	从磁盘进行安装时需要相应的 CD 或 DVD 驱动器
显示器	SQL Server 2008 图形工具需要使用 VGA 或更高分辨率:分辨率至少为 1024×768 像素
其他设备	指针设备:需要 Microsoft 鼠标或兼容的指针设备
操作系统	可以运行在 Windows Vista Home Basic 及更高版本上,也可以在 Windows XP 上运行。从服务器端来看,可以运行在 Windows Server 2003 SP2 及 Windows Server 2008 上,另外,也可以运行在 Windows XP Professional 的 64 位操作系统上以及 Windows Server 2003 和 Windows Server 2008 的 64 位版本上

续表

名称	安装要求
网络组件	IE 浏览器：IE 6.0 SP1 或更高版本 IS：安装报表服务需要 IIS 5.0 以上 ASP.NET 2.0：报表服务需要 ASP.NET
框架	SQL Server 安装程序安装该产品所需的以下软件组件： .NET Framework 3.51 SQL Server Native Client SQL Server 安装程序支持文件
安装软件	SQL Server 安装程序需要使用 Microsoft Windows Installer 4.5 或更高版本以及 Microsoft 数据访问组件(MDAC)2.8 SP1 或更高版本

2.2 MySQL 简介

2.2.1 MySQL

MySQL 是一个精巧的 SQL 数据库管理系统，虽然它不是开放源代码的产品，但在某些情况下可以自由使用。MySQL 的开发者为瑞典 MySQL AB 公司。MySQL 被广泛地应用在 Internet 上的中小型网站中。由于它的强大功能、灵活性、丰富的应用编程接口(API)以及精巧的系统结构，受到了广大自由软件爱好者甚至是商业软件用户的青睐，特别是与 Apache 和 PHP/Perl 结合，为建立基于数据库的动态网站提供了强大动力。

2008 年 1 月 16 号 MySQL AB 被 Sun 公司收购，2010 年 Sun 又被 Oracle 收购，MySQL 成为了 Oracle 公司的另一个数据库项目。

2.2.2 MySQL 经典应用环境

Linux 作为操作系统，Apache 和 Nginx 作为 Web 服务器，MySQL 作为数据库，PHP/Perl/Python 作为服务器端脚本解释器。由于这四个软件都是免费或开放源码软件(FLOSS)，因此使用这种方式不用花一分钱(除开人工成本)就可以建立起一个稳定、免费的网站系统，被业界称为 LAMP 组合。

2.2.3 MySQL 特点

MySQL 拥有很多特点，主要特点如下：

(1) 使用 C 和 C++ 编写，并使用了多种编译器进行测试，保证源代码的可移植性。

(2) 支持 AIX、FreeBSD、HP-UX、Linux、Mac OS、NovellNetware、OpenBSD、OS/2 Wrap、Solaris、Windows 等多种操作系统。

(3) 为多种编程语言提供了 API。这些编程语言包括 C、C++、Python、Java、Perl、PHP、Eiffel、Ruby 和 Tcl 等。

(4) 支持多线程，充分利用 CPU 资源。

（5）优化的 SQL 查询算法，有效地提高查询速度。

（6）既能够作为一个单独的应用程序应用在客户端服务器网络环境中，也能够作为一个库而嵌入其他的软件中。

（7）提供多语言支持，常见的编码如中文的 GB 2312、BIG5、日文的 Shift_JIS 等都可以用作数据表名和数据列名。

（8）提供 TCP/IP、ODBC 和 JDBC 等多种数据库连接途径。

（9）提供用于管理、检查、优化数据库操作的管理工具。

（10）支持大型的数据库。可以处理拥有上千万条记录的大型数据库。可以保存超过 50 000 000 条记录。

（11）支持多种存储引擎。

2.2.4 MySQL 存储引擎

（1）MyISAM：MySQL 5.5 之前的默认数据库引擎，最为常用。拥有较高的插入、查询速度，但不支持事务。

（2）InnoDB：事务型数据库的首选引擎，支持 ACID 事务，支持行级锁定，MySQL 5.5 起成为默认数据库引擎。

（3）BDB：源自 Berkeley DB，事务型数据库的另一种选择，支持 COMMIT 和 ROLLBACK 等其他事务特性。

（4）Memory：所有数据置于内存的存储引擎，拥有极高的插入、更新和查询效率。但是会占用和数据量成正比的内存空间，并且其内容会在 MySQL 重新启动时丢失。

（5）Merge：将一定数量的 MyISAM 表联合而成一个整体，在超大规模数据存储时很有用。

（6）Archive：非常适合存储大量的、独立的、作为历史记录的数据。因为它们不经常被读取。Archive 拥有高效的插入速度，但其对查询的支持相对较差。

（7）Federated 将不同的 MySQL 服务器联合起来，逻辑上组成一个完整的数据库。非常适合分布式应用。

（8）Cluster/NDB 高冗余的存储引擎，用多台数据机器联合提供服务以提高整体性能和安全性。适合数据量大、安全和性能要求高的应用。

（9）CSV：逻辑上由逗号分割数据的存储引擎。它会在数据库子目录里为每个数据表创建一个 .CSV 文件。这是一种普通文本文件，每个数据行占用一个文本行。CSV 存储引擎不支持索引。

（10）BlackHole：黑洞引擎，写入的任何数据都会消失，一般用于记录 binlog 做复制的中继。

（11）EXAMPLE：存储引擎是一个不做任何事情的存根引擎。它的目的是作为 MySQL 源代码中的一个例子，用来演示如何开始编写一个新存储引擎。同样，它的主要兴趣是对开发者。EXAMPLE 存储引擎不支持编索引。

另外，MySQL 的存储引擎接口定义良好。有兴趣的开发者可以通过阅读文档编写自己的存储引擎。

2.2.5 MySQL 应用架构

MySQL 有多种架构方式,不同的架构方式适合不同的应用场合。

单点(single)架构方式,适合小规模应用,其架构如图 2.1 所示。

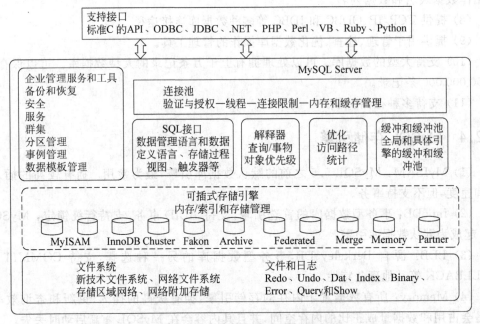

图 2.1 适合小规模应用的 MySQL 架构

复制(replication)架构方式,该方式适合中小规模应用,其架构如图 2.2 所示。

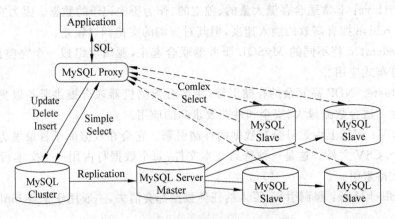

图 2.2 适合中小规模应用 MySQL 架构

集群(cluster)架构方式,适合大规模应用,其架构方式如图 2.3 所示。

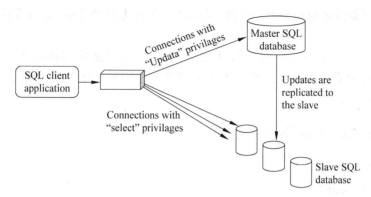

图 2.3　适合大规模应用 MySQL 架构

2.3　Java 及 Java 开发环境简介

2.3.1　Java 起源

Java 是由 Sun Microsystems 公司于 1995 年 5 月推出的 Java 面向对象程序设计语言和 Java 平台的总称。由 James Gosling 和同事们共同研发,并在 1995 年正式推出。用 Java 实现的 HotJava 浏览器(支持 Java applet)显示了 Java 的魅力:跨平台、动态的 Web、Internet 计算。从此,Java 被广泛接受并推动了 Web 的迅速发展,常用的浏览器均支持 Java Applet。另一方面,Java 技术也不断更新。2010 年 Oracle 公司收购了 Sun,Oracle 公司成为 Java 的新主人。

Java 是印度尼西亚爪哇岛的英文名称,因盛产咖啡而闻名。Java 语言中的许多库类名称,多与咖啡有关:如 JavaBeans(咖啡豆)、NetBeans(网络豆)以及 ObjectBeans(对象豆),等等。Sun 和 Java 的标识也正是一杯冒着热气的咖啡。

Java 不同于一般的编译执行计算机语言和解释执行计算机语言,它首先将源代码编译成二进制字节码(byte code),然后依赖各种不同平台上的虚拟机来解释执行字节码。从而实现了"一次编译、到处执行"的跨平台特性。

Java 的诞生是对传统计算机模式的挑战,对计算机软件开发和软件产业都产生了深远的影响:

(1) 软件 4A 目标要求软件能达到任何人在任何地方、任何时间对任何电子设备都能应用。这样能满足软件平台上互相操作,具有可伸缩性和重用性并可即插即用等分布式计算模式的需求。

(2) 基于构建开发方法的崛起,引出了 CORBA 国际标准软件体系结构和多层应用体系框架。在此基础上形成了 Java2 平台和 .NET 平台两大派系,推动了整个 IT 业的发展。

(3) 对软件产业和工业企业都产生了深远的影响,软件从以开发为中心转到了以服务为中心。中间提供商、构件提供商、服务器软件以及咨询服务商出现,企业必须重塑自我,B2B 的电子商务将带动整个新经济市场,使企业获得新的价值、新的增长、新的商机、新的管理。

(4) 对软件开发带来了新的革命,重视使用第三方构件集成,利用平台的基础设施服务,实现开发各个阶段的重要技术,重视开发团队的组织和文化理念、协作、创作、责任、诚信是人才的基本素质。

2.3.2 Java 及 Java 平台的组成

Java 由四方面组成:
(1) Java 编程语言;
(2) Java 类文件格式;
(3) Java 虚拟机(Java Virtual Machine,JVM);
(4) Java 应用程序接口(Java Application Programming Interface,Java API)。

Java 平台由 Java 虚拟机和 Java 应用编程接口构成。Java 应用编程接口为 Java 应用提供了一个独立于操作系统的标准接口,可分为基本部分和扩展部分。

Java 虚拟机是一个想象中的机器,在实际的计算机上通过软件模拟来实现。Java 虚拟机有自己想象中的硬件,如处理器、堆栈、寄存器等,还具有相应的指令系统。引入 Java 虚拟机后,Java 语言在不同平台上运行时不需要重新编译。Java 语言使用模式 Java 虚拟机屏蔽了与具体平台相关的信息,使得 Java 语言编译程序只需生成在 Java 虚拟机上运行的目标代码(字节码),就可以在多种平台上不加修改地运行。Java 虚拟机在执行字节码时,把字节码解释成具体平台上的机器指令执行。

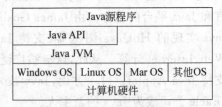

图 2.4 Java 环境

Java 虚拟机、Java 应用编程接口、操作系统和硬件之间的关系如图 2.4 所示。

2.3.3 Java 的版本

目前 Java 主要有三个运行在不同环境上的版本。

(1) J2SE(Java 2 Platform Standard Edition,Java 平台标准版),Java SE(Java Platform Standard Edition)。Java SE 以前称为 J2SE。它允许开发和部署在桌面、服务器、嵌入式环境和实时环境中使用的 Java 应用程序。Java SE 包含了支持 Java Web 服务开发的类,并为 Java Platform Enterprise Edition(Java EE)提供基础。

(2) J2EE(Java 2 Platform Enterprise Edition,Java 平台企业版),Java EE(Java Platform,Enterprise Edition)。这个版本以前称为 J2EE。企业版本帮助开发和部署可移植、健壮、可伸缩且安全的服务器端 Java 应用程序。Java EE 是在 Java SE 的基础上构建的,它提供 Web 服务、组件模型、管理和通信 API,可以用来实现企业级的面向服务体系结构(Service-Oriented Architecture,SOA)和 Web 2.0 应用程序。

(3) J2ME(Java 2 Platform Micro Edition,java 平台微型版),Java ME(Java

Platform Micro Edition)。这个版本以前称为 J2ME,也叫 K-Java。Java ME 为在移动设备和嵌入式设备(例如手机、PDA、电视机顶盒和打印机)上运行的应用程序提供一个健壮且灵活的环境。Java ME 包括灵活的用户界面、健壮的安全模型、许多内置的网络协议以及对可以动态下载的连网和离线应用程序的丰富支持。基于 Java ME 规范的应用程序只需编写一次,就可以用于许多设备,而且可以利用每个设备的本机功能。

2.3.4 Java 的相关技术和主要特性

Sun 公司对 Java 编程语言的解释是:Java 编程语言是简单、面向对象、分布式、解释性、健壮、安全与系统无关、可移植、高性能、多线程和动态的语言。

Java 语言的优良特性使得 Java 应用具有无比的健壮性和可靠性,这也减少了应用系统的维护费用。

与 Java 语言相关的技术非常多,主要的相关技术如下:

(1) JDBC(Java Database Connectivity)提供连接各种关系数据库的统一接口,作为数据源,可以为多种关系数据库提供统一访问,它由一组用 Java 语言编写的类和接口组成。

(2) EJB(Enterprise JavaBeans)使开发者方便地创建、部署和管理跨平台的基于组件的企业应用。

(3) Java RMI(Java Remote Method Invocation)用来开发分布式 Java 应用程序。一个 Java 对象的方法能被远程 Java 虚拟机调用。

(4) Java IDL(Java Interface Definition Language) 提供与 CORBA(Common Object Request Broker Architecture)的无缝的互操作性。

(5) JNDI(Java Naming and Directory Interface)提供从 Java 平台到统一的客户端 API 的无缝连接。这个接口屏蔽了企业网络所使用的各种命名和目录服务。

(6) JMAPI(Java Management API)为异构网络上系统、网络和服务管理的开发提供一整套丰富的对象和方法。

(7) JMS(Java Message Service)提供企业消息服务,如可靠的消息队列、发布和订阅通信以及有关推拉(Push/Pull)技术的各个方面。

(8) JTS(Java transaction Service)提供存取事务处理资源的开放标准,这些事务处理资源包括事务处理应用程序、事务处理管理及监控。

(9) JMF(Java Media Framework API)可以帮助开发者把音频、视频和其他一些基于时间的媒体放到 Java 应用程序或 applet 小程序中去,为多媒体开发者提供了捕捉、回放、编解码等工具,是一个弹性的、跨平台的多媒体解决方案。

(10) JMX(Java Management Extensions,即 Java 管理扩展)是一个为应用程序、设备、系统等植入管理功能的框架。JMX 可以跨越一系列异构操作系统平台、系统体系结构和网络传输协议,灵活地开发无缝集成的系统、网络和服务管理应用。

(11) JSP(Java Server Pages)是由 Sun Microsystems 公司倡导、许多公司参与一起建立的一种动态网页技术标准。JSP 技术有点类似 ASP 技术,它是在传统的网页 HTML 文件(*.htm、*.html)中插入 Java 程序段(Scriptlet)和 JSP 标记(tag),从而形成JSP文件(*.jsp)。用JSP开发的Web应用是跨平台的,既能在Linux下运行,也能在其他操作系统上运行。

Java IDE (Java Integrated Development Environment,Java集成开发环境)当今最流行的是Eclipse、MyEclipse、IntelliJ、JBuilder、JDeveloper、NetBeans、JCreator等。

Java语言的主要特性有以下几方面:

(1) Java语言是简单的。Java语言的语法与C语言和C++语言很接近,使得大多数程序员很容易学习和使用Java。另一方面,Java丢弃了C++中很少使用的、很难理解的、令人迷惑的那些特性,Java语言不使用指针,而是引用,并提供了自动的废料收集,使得程序员不必为内存管理而担忧。

(2) Java语言是面向对象的。

(3) Java语言是分布式的。Java语言支持Internet应用的开发,在基本的Java应用编程接口中有一个网络应用编程接口(Java net),它提供了用于网络应用编程的类库,包括URL、URLConnection、Socket、ServerSocket等。Java的RMI(远程方法激活)机制也是开发分布式应用的重要手段。

(4) Java语言是健壮的。

(5) Java语言是体系结构中立的。

(6) Java语言是可移植的。

(7) Java语言是解释型的。

(8) Java是高性能的。

(9) Java语言是多线程的。Java中的类有一个运行时刻的表示,能进行运行时刻的类型检查。Java语言的优势主要有以下五点:

① 一次编写,到处运行。除了系统之外,代码不用做任何更改。

② 系统的多平台支持。基本上可以在所有平台上的任意环境中开发,在任意环境中进行系统部署,在任意环境中扩展。

③ 强大的可伸缩性。从只有一个小的Jar文件(Java Archive,Java归档文件)就可以运行Servlet/JSP,到由多台服务器进行集群和负载均衡,到多台Application进行事务处理,消息处理,从一台服务器到无数台服务器,Java显示了巨大的生命力。

④ 多样化和功能强大的开发工具支持。Java已经有了许多非常优秀的开发工具,而且许多可以免费得到,并且其中许多已经可以顺利地运行于多种平台之下。

⑤ 支持服务器端组件。JSP可以使用成熟的JavaBeans组件实现复杂商务功能。

Java的一些优势正是它致命的问题所在,其弱势在于:

(1) 正是由于跨平台的功能,为了实现极度的伸缩能力,所以极大地增加了产品的复杂性。

(2) Java的运行速度是用class常驻内存来完成的,所以它在一些情况下所使用的内存比起用户数量来说确实是"最低性能价格比"了。

2.3.5 JSP简介

JSP是由Sun Microsystems公司倡导、许多公司参与一起建立的一种动态技术标准。在传统的网页 HTML 文件(*.htm、*.html)中加入 Java 程序片段(Scriptlet)和 JSP 标签,就构成了 JSP 网页。Java 程序片段可以操纵数据库、重新定向网页以及发送 E-mail 等,实现建立动态网站所需要的功能。所有程序操作都在服务器端执行,网络上传送给客户端的仅是得到的结果,这样大大降低了对客户浏览器的要求,即使客户浏览器端不支持 Java,也可以访问 JSP 网页。

简单说 JSP(Java Server Pages)是一种跨平台的动态网页技术,在静态页面中嵌入 Java 代码片段,再由 Web 服务器中的 JSP 引擎进行编译并执行嵌入的 Java 代码片段,生成的页面信息返回给客户端。

自 JSP 推出后,众多大公司都支持 JSP 技术的服务器,如 IBM、Oracle、Bea 公司等,所以 JSP 迅速成为商业应用的服务器端语言。

一个 JSP 网页面可以被分为以下几部分:

(1) 静态数据,如 HTML;

(2) JSP 指令,如 include 指令;

(3) JSP 脚本元素和变量;

(4) JSP 动作;

(5) 用户自定义标签。

一个 JSP 页面有多个客户访问,下面是第一个客户访问 JSP 页面时候,JSP 页面的执行流程:

(1) 客户通过浏览器向服务器端的 JSP 页面发送请求。

(2) JSP 引擎检查 JSP 文件对应的 Servlet 源代码是否存在,若不存在转向第(4)步,否则执行下一步。

(3) JSP 引擎检查 JSP 页面是否需要修改,若没修改,转向第(5)步,否则执行下一步。

(4) JSP 引擎将 JSP 页面文件转译为 Servlet 源代码(相应的 .java 代码)。

(5) JSP 引擎将 Servlet 源代码编译为相应字节码(.class 代码)。

(6) JSP 引擎加载字节码到内存。

(7) 字节码处理客户请求,并将结果返回给客户。

JSP 网页的执行过程如图 2.5 所示。

在不修改 JSP 页面的情况下,除了第一个客户访问 JSP 页面需要经过以上几个步骤外,以后访问该 JSP 页面的客户请求,直接发送给 JSP 对应的字节码程序处理,并将处理结果返回给客户。

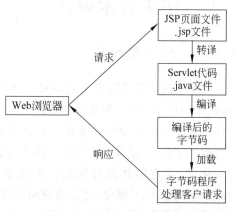

图 2.5 JSP 网页的执行过程

2.3.6 Java 的开发环境 MyEclipse 8.6 简介

MyEclipse 是 Java 众多集成开发环境中的流行最广者之一,MyEclipse 企业级工作平台(MyEclipse Enterprise Workbench,简称 MyEclipse)是对 Eclipse IDE 的扩展,利用它可以在数据库和 Java EE 的开发、发布以及应用程序服务器的整合方面极大地提高工作效率。它是功能丰富的 Java EE 集成开发环境,包括完备的编码、调试、测试和发布功能,完整支持 HTML、Struts、JSP、CSS、JavaScript、Spring 以及 SQL。

与早版本的 Eclipse 相比,MyEclipse 8.6 的主要改进包括:

(1) 引入了一个全新的 JavaScript 编辑器,该编辑器提供了更好的 JavaScript/HTML 高亮标记和代码支持,同时提供了更为精确的语法校验功能。

(2) 新增了 Struts2 图形编辑器(是 MyEclipse 对 Struts 支持的一个卖点)。

(3) 引入了 MyEclipse 配置中心功能(该功能是 MyEclipse 8.6 中的重大卖点)。

(4) 该功能包括允许用户更方便的安装/卸载 MyEclipse 模块。

(5) 快速检索和安装目前最流行的 Eclipse 插件。

(6) 浏览、编辑和安装所有的软件更新(同时支持自定义升级站点)。

(7) 轻松地在团队中共享 MyEclipse 8.6 工作平台配置。

(8) 通过授权获取用户的工作台变更信息,同时让用户共享这些工作台的配置。

(9) 通过 MyEclipse 配置中心持久化(保存)工作台的个性化设置。

(10) 允许独立共享工作台配置信息,也可以通过授权让特定的用户更改工作台配置信息。

(11) 新增了大量的应用程序服务器连接器,更多地应用程序服务器。

2.4 .NET 技术简介

2.4.1 .NET 是什么

.NET 就是微软用来实现 XML、Web Services、SOA(面向服务的体系结构 service-oriented architecture)和敏捷性的技术。对技术人员,微软搭建技术平台,技术人员在这个技术平台之上创建应用系统。从这个角度,.NET 也可以定义为:.NET 是微软的新一代技术平台,为敏捷商务构建互联互通的应用系统,这些系统是基于标准的、联通的、适应变化的、稳定的和高性能的。从技术的角度,一个.NET 应用是一个运行于.NET Framework(.NET 框架)之上的应用程序。换句话说,一个.NET 应用是一个使用.NET 框架类库来编写,并运行于公共语言运行时(Common Language Runtime)之上的应用程序。如果一个应用程序跟.NET 框架无关,它就不能叫做.NET 程序。

.NET 是以 Internet 为中心的一种全新的开发平台;通过.NET,可以将用户数据存放在网络上,并且随时随地通过与.NET 兼容的任何设备访问这些数据;.NET 独一无二的特征是可以提供多语言支持;.NET 平台框架开发出来的程序,可以在不同的平台上运行,实现一次编写,到处运行;.NET 是一个环境,是一个互联网的开发平台..NET 是 Microsoft XML Web Services 平台。XML Web Services 允许应用程序通过 Internet 进行通信和共享数据,而不管所采用的是哪种操作系统、设备或编程语言。

将.NET 分成以下四个主要部分。

(1).NET 战略：该战略是基于这样一种想法,即所有的设备将来会通过一个全球宽带网(即 Internet)连接在一起,这个软件就成为在该网络上提供的一种服务。

(2).NET Framework：是指像 ASP.NET 这样可使.NET 更加具体的新技术。该架构提供了具体的服务和技术,以便于开发人员创建应用程序,以满足如今连接到 Internet 上的用户的需要。

(3) Windows 服务器系统：是指像 SQL Server 2000 和 BizTalk Server 2000 这样的由.NET Framework 应用程序使用的服务器产品,不过目前它们并不是使用.NET Framework 编写的。这些服务器产品将来的版本都将支持.NET,但不必使用.NET 重新编写。

(4) 对开发人员来说,.NET 平台中另一个重要的部分自然就是开发工具了。为此 Microsoft 公司对.NET 的首要开发环境 Visual Studio 进行了最新一次的重要升级,该升级产品称为 Visual Studio.NET。不过,仍然可以使用 Notepad 或其他 IDE 来开发.NET 应用程序,这也是 Microsoft 公司中许多开发小组所采用的方式。

2.4.2 .NET 框架

.NET 框架是.NET 开发、运行的基础架构,是一种新的计算平台,它简化了在高度分布式 Internet 环境中的应用程序开发。

.NET 框架具有两个主要组件。

(1) 公共语言运行库；公共语言运行库是.NET 框架的基础。用户可以将运行库看作一个在执行时管理代码的代理,它提供核心服务(如内存管理、线程管理和远程处理),而且还强制实施严格的类型安全以及可确保安全性和可靠性的其他形式的代码准确性。事实上,代码管理的概念是运行库的基本原则。以运行库为目标的代码称为托管代码,而不以运行库为目标的代码称为非托管代码。

(2).NET 框架类库。类库是一个综合性的面向对象的可重用类型集合,可以使用它开发包含从传统的命令行或图形用户界面(GUI)应用程序到基于 ASP.NET 所提供的最新创新的应用程序(如 Web 窗体和 XML Web Services)在内的应用程序。

.NET 框架设计为一个集成环境,.NET 框架旨在实现下列目标：

提供一个覆盖整个应用范围的、一致的面向对象环境；对象代码是在本地存储和执行,还是在本地执行但在 Internet 上分布,或者是在远程执行的。提供一个将软件部署和版本控制冲突最小化的代码执行环境。提供一个保证代码(包括由未知的或不完全受信任的第三方创建的代码)安全执行的代码执行环境。提供一个可消除脚本环境或解释环境的性能问题的代码执行环境。使开发人员的经验在面对类型大不相同的应用程序(如基于 Windows 的应用程序和基于 Web 的应用程序)时保持一致。

为了实现上述目标,.NET 框架设计者们最后确定了以下体系结构,将框架分解为两部分：通用语言运行时 CLR 和框架类库 FCL,其结构如图 2.6 所示。

CLR 是 Microsoft 对 CLI 标准(Common Language Infrastructure,公共语言基础结构,已经得到了 ECMA(欧洲计算机制造协会)的标准化)的具体实现,它处理代码执行及所有相关任务：编译、内存管理、安全、线程管理、强制类型安全和类型使用。在 CLR 中运行的代码称为托管代码(managed code),以区别于不在 CLR 中运行的非托管代码(unmanaged code),如基于 COM 或 Windows API 的组件。图 2.7 给出了 CLI 规范定义的框架示意。

图 2.6 .NET 框架示意图

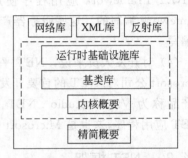

图 2.7 CLI 规范定义的框架

.NET 的另一个主要部分是框架类库 FCL,对于在.NET 中运行的应用来说,它是一个可重用的类型(类、结构等)代码库。只要是遵循.NET 框架的语言,都会使用这个公共类库。因此,只要知道了如何使用这些类型,不论选择用哪一种.NET 语言编写程序,这些类都可以使用。

.NET 框架可由非托管组件承载,这些组件将公共语言运行库加载到它们的进程中并启动托管代码的执行,从而创建一个可以同时利用托管和非托管功能的软件环境。.NET 框架不但提供若干个运行库宿主,而且还支持第三方运行库宿主的开发。

2.4.3 .NET 的特点

.NET 具有以下特点：

(1) 基于组件的技术。.NET 组件不需要写额外的底层代码来支持组件化,只需一个.NET 类,不使用注册表。

(2) 跨语言集成。目前支持.NET 平台开发的编程语言超过了 25 种;.NET 编程语言的跨语言集成是在 IL 层次上面集成,可以调式抛出的异常,可以扩展相应的功能。

(3) 简化开发。.NET 提供一套框架类,允许任何语言使用。每次更换语言时无须学习新 API;Visual Studio.NET 中不同语言所对应的开发环境一模一样,提高开发效率。

(4) 简化部署。.NET 环境中可以执行文件使用共享 DLL,.NET 去掉了注册设置;.V 引入"安装卸载零影响"的概念。在.NET 中安装文件只需要从 CD 上的一个目录拷贝到另一个目录就行了。

（5）强大的分布式应用。.NET 技术采用 XML 的编码风格，应用 SOAP 协议进行分布式的调用。

（6）可靠性和安全性。.NET 程序在编译和运行中借助 CLR 托管执行；类型的安全检查、垃圾收集、即时编译等；声明性的安全检查、强制性的安全检查。

在开发技术方面，.NET 提供了全新的数据库访问技术 ADO.NET，以及网络应用开发技术 ASP.NET 和 Windows 编程技术 Win Forms；在开发语言方面，.NET 提供了 VB、VC++、C♯ 和 JScript 等多种语言支持；而 Visual Studio.NET 则是全面支持.NET 的开发工具。

.NET 所支持的语言有 C♯、Visual J♯、COBOL、VB.NET、VC++.NET、SmallTalk、JScript.NET 和 ASP.NET 等。

2.4.4 .NET 的版本

从 2002 年 2 月以来，已经推出了七个主要版本。这些版本的主要情况如表 2.4 所示。

表 2.4 不同.NET 版本简单说明

版本	完整版本号	发行日期	Visual Studio	Windows 默认安装
1.0	1.0.3705.0	2002-02-13	Visual Studio.NET 2002	Windows XP Media Center Edition Windows XP Tablet PC Edition
1.1	1.1.4322.573	2003-04-24	Visual Studio.NET 2003	Windows Server 2003
2.0	2.0.50727.42	2005-11-07	Visual Studio 2005	
3.0	3.0.4506.30	2006-11-06		Windows Vista Windows Server 2008
3.5	3.5.21022.8	2007-11-19	Visual Studio 2008	Windows 7 Windows Server 2008 R2
4.0	4.0.30319.1	2010-04-12	Visual Studio 2010	
4.5	4.5.40805	2012-02-20	Visual Studio 2012 RC	Windows 8 RP Windows Server 8 RC

2.5 三层架构和 MVC 架构简介

2.5.1 三层架构简介

三层架构（3-tier application）就是将整个业务应用划分为表现层（User Interface，UI）、业务逻辑层（Business Logic Layer，BLL）、数据访问层（Data Access Layer，DAL）。区分层次的目的是为了"高内聚，低耦合"。

表现层就是展现给用户的界面，即用户在使用一个系统的时候所见所得。业务逻辑层就是针对具体问题的操作,也可以说是对数据层的操作,对数据业务逻辑处理。数据访问层是直接操作数据库,针对数据的增添、删除、修改、更新、查找等。三层架构形象地表示如图2.8所示。

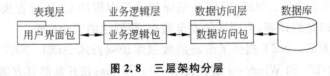

图2.8 三层架构分层

三个层次中,系统主要功能和业务逻辑都在业务逻辑层进行处理。

三层体系的应用程序将业务规则、数据访问、合法性校验等工作放到了中间层(业务逻辑层)进行处理。通常情况下,客户端不直接与数据库进行交互,而是通过COM/DCOM通信与中间层建立连接,再经由中间层与数据库进行交互。

位于最外层的表现层,离用户最近。它用于显示数据和接收用户输入的数据,为用户提供一种交互式操作的界面。

位于中间的业务逻辑层是架构中体现核心价值的部分。它的关注点主要集中在业务规则的制定、业务流程的实现等与业务需求有关的系统设计,也即是说它是与系统所应对的领域逻辑有关,很多时候,也将业务逻辑层称为领域层。业务逻辑层在体系架构中的位置很关键,它处于数据访问层与表现层中间,起到了数据交换中承上启下的作用。由于层是一种弱耦合结构,层与层之间的依赖是向下的,底层对于上层而言是"无知"的,改变上层的设计对于其调用的底层而言没有任何影响。如果在分层设计时,遵循了面向接口设计的思想,那么这种向下的依赖也应该是一种弱依赖关系。因而在不改变接口定义的前提下,理想的分层式架构应该是一个支持可抽取、可替换的"抽屉"式架构。正因为如此,业务逻辑层的设计对于一个支持可扩展的架构尤为关键,因为它扮演了两个不同的角色。对于数据访问层而言,它是调用者;对于表现层而言,它却是被调用者。依赖与被依赖的关系都纠结在业务逻辑层上,如何实现依赖关系的解耦,则是除了实现业务逻辑之外留给设计师的任务。

数据访问层有时候也称为是持久层,其功能主要是负责数据库的访问,可以访问数据库系统、二进制文件、文本文档或是XML文档。简单说就是实现对数据表的Select、Insert、Update、Delete的操作。

三层架构将数据层、应用层和业务层分离,业务层通过应用层访问数据库,保护数据安全,利于负载平衡,提高运行效率,方便构建不同网络环境下的分布式应用。

同时三层架构提供了非常好的可扩张性,可以将逻辑服务分布到多台服务器来处理,从而提供了良好的伸缩方案。

三层架构优点主要体现在以下几方面:

(1) 开发人员可以只关注整个架构中的其中某一层;

(2) 可以很容易的用新的实现来替换原有层次的实现;

(3) 可以降低层与层之间的依赖;

(4) 有利于标准化;

(5)利于各层逻辑的复用。

三层架构也有其问题,主要缺点有:

(1)降低了系统的性能。如果不采用分层式架构,很多业务可以直接造访数据库,以此获取相应的数据,如今却必须通过中间层来完成。

(2)有时会导致级联的修改。这种修改尤其体现在自上而下的方向。如果在表现层中需要增加一个功能,为保证其设计符合分层式结构,可能需要在相应的业务逻辑层和数据访问层中都增加相应的代码。

2.5.2 MVC框架简介

MVC是model(模型)- view(视图)- controller(控制器)的缩写,是XeroxPARC在20世纪80年代为编程语言Smalltalk-80发明的一种软件设计模式,至今已被广泛使用。最近几年被推荐为Sun公司J2EE平台的设计模式,并且受到越来越多的使用ColdFusion和PHP的开发者的欢迎。模型—视图—控制器模式是一个有用的工具箱。

MVC是一种软件设计典范,用于组织代码的一种业务逻辑和数据显示分离的方法,其核心思想是业务逻辑被聚集到一个部件里面,当界面和用户围绕数据的交互能被改进和个性化定制时不需要重新编写业务逻辑;使用MVC的目的是将M和V的实现代码分离,从而把同一个程序的处理结果以不同的形式表现出来,C存在的目的则是确保M和V的同步,一旦M改变,V应该同步更新。MVC的框架结构如图2.9所示。

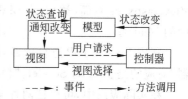

图2.9 MVC框架结构

视图是用户看到并与之交互的界面,通常是Web页面。在视图中没有真正的处理发生,其功能是解释模型、模型更新请求、发送用户输入给控制器、允许控制器选择视图。

模型是用来表示数据和业务的规则。在MVC的三个部件中,模型拥有最多的处理任务,其主要功能是封装应用程序状态、响应状态查询、应用程序功能、通知视图改变。被模型返回的数据是中立的,就是说模型与数据格式无关,这样就使得一个模型能为多个视图提供数据。

控制器接受用户的输入并调用模型和视图去完成用户的需求,其主要功能是定义引用程序行为、用户动作映射成模型更新、选择相应的视图等。当单击Web页面中的超链接和发送HTML表单时,控制器本身不输出任何东西也不做任何处理,它只是接收请求并决定调用哪个模型构件去处理请求,然后确定用哪个视图来显示模型处理返回的数据。

MVC架构最大优势之一是它能为应用程序处理很多不同的视图。由于应用于模型的代码只需写一次就可以被多个视图重用,所以减少了代码的重复性。

因为模型是自包含的,并且与控制器和视图相分离,所以很容易改变你的应用程序的数据层和业务规则。

由于可以使用控制器来连接不同的模型和视图去完成用户的需求,这样控制器可以为构造应用程序提供强有力的手段。给定一些可重用的模型和视图,控制器可以根据用户的需求选择模型进行处理,然后选择视图将处理结果显示给用户。

由于运用 MVC 的应用程序的三个部件是相互独立,改变其中一个不会影响其他两个,所以依据这种设计思想能构造良好的松耦合的构件。

由于 MVC 没有明确的定义,它的内部原理比较复杂,应用时有一定难度。

2.5.3 三层架构和 MVC 框架的关系

MVC 设计模式是从早期的客户/服务器应用发展而来的,它采用的是两层架构设计,由于三层架构是对两层架构的延伸,它们之间的关系如图 2.10 所示。

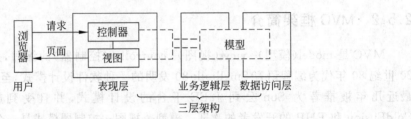

图 2.10 MVC 与三层架构的关系

利用.NET 进行项目开发时,三层架构中"表现层"的 aspx 页面对应 MVC 中的 View,aspx.cs 页面(类)对应 MVC 中的 Controller,业务逻辑层和数据访问层对应 MVC 中的 Model。

利用 Java 和 JSP 进行项目开发时,MVC 框架中的"视图"对应的是.jsp 文件,"控制器"对应的是 Java 的 Servlet,"模型"对应的是 Java 的 Bean。

在实际应用中,可以把两种架构结合在一起,其结构如图 2.11 所示。

图 2.11 两种模式结合成的结构

第 3 章 ATM 存取款管理系统设计与实现

本章介绍一个简单的 ATM 存取款管理系统的设计与实现,应用 Java 编程语言和面向对象开发方法完成,目的是让大家熟悉 Java 面向对象基本特点——继承、封装和多态以及异常处理、输入输出——在编程中的实现,为以后进行较复杂的项目开发打下良好基础。该系统只关注如何用面向对象的思维和方法将问题描述清楚,不要求实现 GUI,只考虑系统逻辑实现。

3.1 项目需求分析

为了使程序实现更加简单,更好地理解面向对象编程思想,本系统模拟一般的 ATM 机功能并做了简化,只针对其主要业务加以定义和实现。

ATM 系统的基本功能包括用户身份验证、存款、取款、查询余额、修改密码、查看用户信息。具体需求如下:

(1) 用户身份验证。进入 ATM 管理系统之前,输入用户账号和密码,模拟刷卡输入密码的过程,以验证用户身份是否合法。

(2) 存款。系统主要功能之一,需要提示用户存款额和账户余额数。

(3) 取款。系统主要功能之一,要验证取款额不能大于账户余额,并提示用户取款额和账户余额数。

(4) 查询余额。查询当前账户的账面余额。

(5) 修改密码。提示输入用户输入原密码,验证通过后两次输入新密码,如果两次输入密码不一致,提示密码修改失败;原密码输错三次则提示本次不能再修改密码了。

(6) 查看用户信息。显示当前账户的账号、用户名、年龄、账户余额等信息。

3.2 面向对象的分析与设计

3.2.1 实体类分析与设计

面向对象设计把握一个重要的原则:谁拥有数据,谁就对外提供操作这些数据的方法。

ATM 管理系统中,有几种实体数据是必须要表达的,如账户数据、用户数据、银行卡数据。因此,可以设计以下三个类:

(1) 银行卡类 Card。封装银行卡号 cardNumbr 和对银行卡号进行操作的 get、set 方法。

(2) 用户类 User。封装用户姓名、性别、年龄和所持有的银行卡等信息,并通过 get、set 方法对这些信息进行操作。

(3) 账户类 Account。封装账户密码、账户余额和账户所关联的用户等信息,并在这个类中提供存钱、取钱、查询余额、修改密码等业务方法。

各个类的图示如图 3.1 所示。

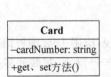  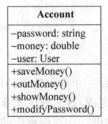

图 3.1 ATM 管理系统实体类图

3.2.2 工具类分析与设计

而系统的处理过程中,有很多具有共性的输入输出功能,可以将其定义成静态方法,全部封装在一个工具类 Tools 里,方便以后调用。类图如图 3.2 所示。

其中三个 input 方法分别通过键盘读取用户输入的 double 型、String 型和 int 型数据,在 ATM 系统中多处需要从键盘输入数据,如输入账号、输入密码、输入存取款钱数……此时,调用这些静态方法将大大提高代码的复用性,而且使程序更简洁,结构性更好。

图 3.2 ATM 管理系统工具类图

而三个重载的 show 方法可以根据参数的不同分别输出银行卡信息、用户信息和银行账户信息,即 show(Card) 输出银行卡信息,show(User) 输出用户信息,而 show(Account) 输出账户信息。这是 Java 面向对象多态性的一个很好应用。

最后,可以将显示 ATM 主界面信息的功能定义在 showATMInfo() 方法中从而简化后面主类的设计,使主类的功能更单一。

3.2.3 主类分析与设计

最后,设计主类 MainClass 实现系统业务流程的控制,只在主类中定义主方法实现系统的流程控制。

ATM 管理系统流程图如图 3.3 所示。

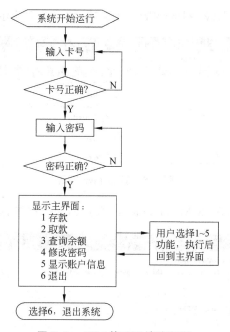

图 3.3　ATM 管理系统流程图

3.3　系统实现与测试

3.3.1　项目环境准备

我们在 MyEclipse 8.6 开发环境下完成这个项目(其他任何环境下也都可以顺利完成),下面以 MyEclipse 8.6 开发环境为例逐步完成项目的构建。

启动 MyEclipse,主体界面如图 3.4 所示。

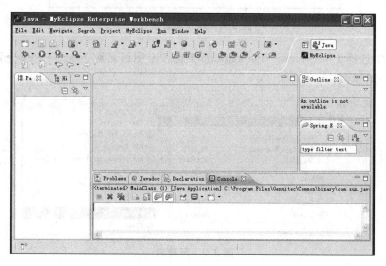

图 3.4　MyEclipse 主界面

在菜单中选择 File→New→Java Project 选项,如图 3.5 所示。

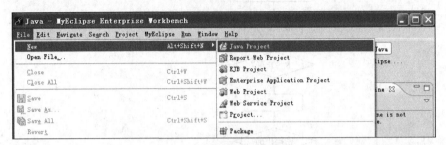

图 3.5 新建 Java 项目

输入项目名称,例如,ATMManagement,单击 Finish 按钮,如图 3.6 所示。

图 3.6 输入项目信息

在菜单中选择 File→New→Package 选项或右击项目名称下 src→New→Package(如图 3.7 所示)新建包。

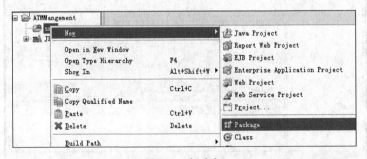

图 3.7 新建包

输入包名 myproject.atm 并单击 Finish 按钮,如图 3.8 所示。

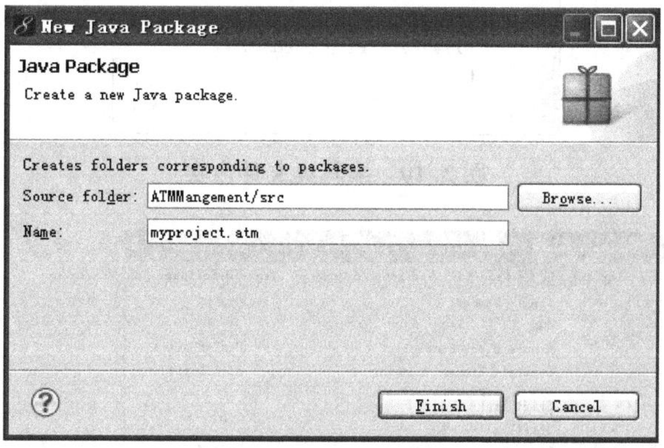

图 3.8　输入包名

在 Java 中,不推荐把类放在无名包中,所以项目开发时,一般都会按照功能相似或性质相同的规律建立几个包,把类放到相应包下。

因为我们的项目比较简单,只定义一个包 myproject.atm,将所有类定义在这个包下。

在菜单中选择 File→New→Class 新建类,如图 3.9 所示。

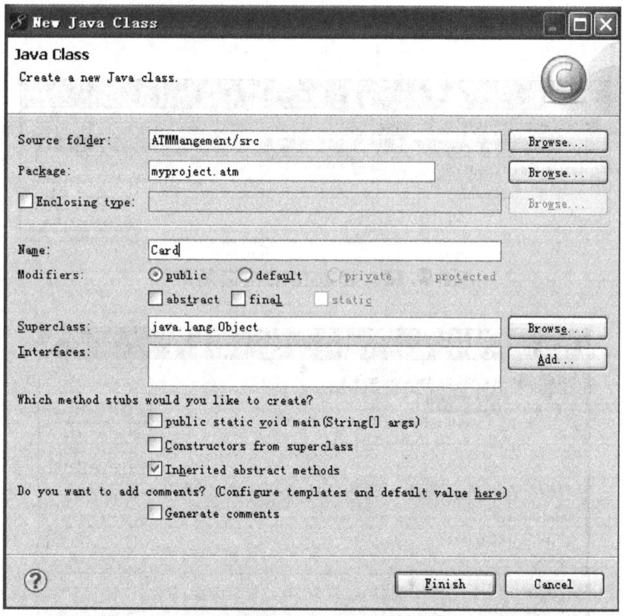

图 3.9　新建类

单击 Finish 按钮,打开 Card.java 文件编辑页面(如图 3.10 所示),输入类文件内容。

注意:在 MyEclipse 中,可以借助软件自带的功能自动生成构造方法和 get、set 方法,大大简化代码的编写,提高编码速度,如图 3.11～图 3.13 所示。

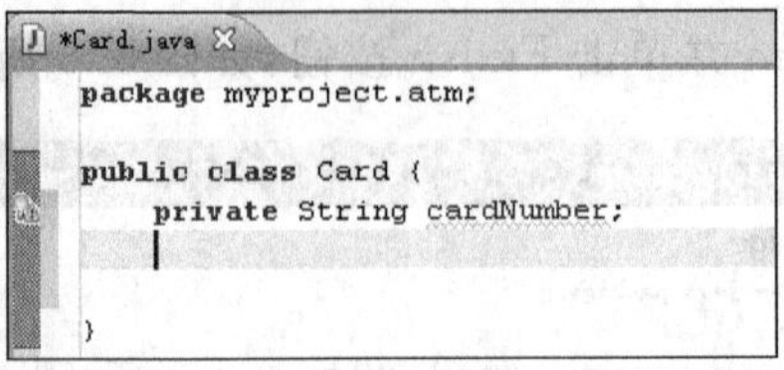

58 \软\件\项\目\实\践\案\例\教\程\

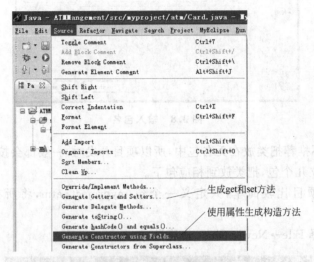

图 3.10　编写类文件内容

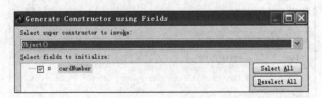

图 3.11　自动生成代码

图 3.12　自动生成构造函数

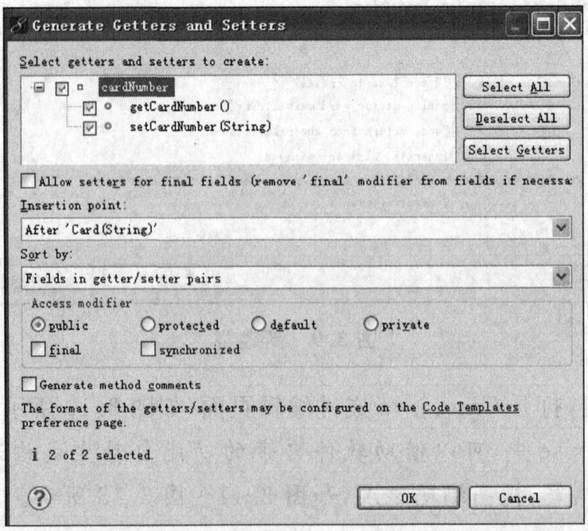

图 3.13　自动生成 get、set 方法

在 Card 类中编写主方法，测试一下，如图 3.14 所示。

```
*Card.java
package myproject.atm;
public class Card {
    private String cardNumber;
    public Card() {
        super();
    }
    public Card(String cardNumber) {
        super();
        this.cardNumber = cardNumber;
    }
    public String getCardNumber() {
        return cardNumber;
    }
    public void setCardNumber(String cardNumber) {
        this.cardNumber = cardNumber;
    }
    //编写主方法测试一下
    public static void main(String args[]){
        Card card=new Card("62251000");
        System.out.println("卡号是: "+card.getCardNumber());
        card.setCardNumber("62252000");
        System.out.println("改变后卡号是: "+card.getCardNumber());
    }
```

图 3.14　编写主方法测试

执行 Run→Run 命令或者按下 Ctrl＋F11 键，运行文件，在控制台 Console 窗口看到的运行结果如图 3.15 所示。

```
Problems  @ Javadoc  Declaration  Console
<terminated> Card [Java Application] C:\Program Files\Genuitec\Common\b
卡号是: 62251000
改变后卡号是: 62252000
```

图 3.15　运行结果

3.3.2　项目类定义与实现

在项目环境下依次实现三个实体类 Card、User 和 Account，工具类 Tods 以及主类 MainClass 的定义，参考代码如下。

1. 银行卡类源程序 Card.java

略。参见图 3.14。

2. 用户信息类源程序 User.java

```
package myproject.atm;

public class User {
    //属性
    private String username=null;
    private char sex='男';
    private int age=0;
    private Card card=new Card();
```

```java
    //构造方法
    public User(String username, char sex, int age, Card card){
        //略,请自动生成
    }
    public User(){
        super();
    }
    //get、set 成员方法略,请自动生成
}
```

3. 账户类源程序 Account.java

```java
package myproject.atm;
public class Account {
    //封装私有属性
    private String password=null;
    private double money=0.0;
    private User user=new User();

    //构造方法略,请自动生成
    //get、set 成员方法略,请自动生成

    //存钱方法
    public void saveMoney(){
        System.out.print("请输入你的存款金额:");
        double money=0.0;
        money=Tools.inputDouble();              //从键盘上读取一个实数
        this.money+=money;
        System.out.println("您本次存入:"+money+"元,账户余额是:"+this.money+"元");
    }
    //取钱方法
    public void outMoney(){
        double money=0.0;
        System.out.print("请输入取款金额:");
        money=Tools.inputDouble();              //从键盘上读取一个实数
        if(money>this.money){
            System.out.print("余额不足!");
        }else{
            this.money-=money;
            System.out.println("您本次取款"+money+"元,账户余额:"+this.money);
        }
    }
    //显示账户余额
    public void showMoney(){
        System.out.println("您的账户余额是:"+"  "+money+"  元");
```

```java
        }
        //修改密码
        public void modifyPassword(){
            String s1=null, s2=null;
            int num=3;
            System.out.print("请输入原密码:");
            while(num-->0){
                if(Tools.inputString().equals(this.password)){
                    System.out.print("请输入新的密码:");
                    s1=Tools.inputString();
                    System.out.print("请再输入一次:");
                    s2=Tools.inputString();
                    if(s1.equals(s2)){
                        password=s1;
                        System.out.println("修改密码成功~");
                    }else{
                        System.out.println("两次输入的密码不一致,密码修改失败!");
                    }
                    return;
                }else{
                    if(num==0){
                        System.out.println("您已输错 3 次密码,本次不再允许修改~");
                        break;
                    }
                    System.out.print("密码不对,请重新输入:");
                }
            }
        }
    }
```

4．工具类源程序 Tools.java

```java
package myproject.atm;
import java.util.Scanner;
public class Tools {
    //显示银行卡信息
    public static void show(Card card){
        System.out.println("卡号:"+card.getCardNumber());
    }
    //显示用户信息
    public static void show(User user){
        show(user.getCard());
        System.out.println("姓名:"+user.getUsername());
        System.out.println("性别:"+user.getSex());
        System.out.println("年龄:"+user.getAge());
```

```java
    }
    //显示账户信息
    public static void show(Account acount){
        show(acount.getUser());
        System.out.println("密码:"+acount.getPassword());
        System.out.println("余额: "+acount.getMoney());
    }
    //显示提示用户操作的主界面信息
    public static void showATMInfo(){
        System.out.println("ATM存取款管理系统,请选择你的操作:");
        System.out.println("--------------------------------");
        System.out.println("   1   存款");
        System.out.println("   2   取款");
        System.out.println("   3   查询余额");
        System.out.println("   4   修改密码 ");
        System.out.println("   5   显示账户信息");
        System.out.println("   6   退出       ");
        System.out.println("--------------------------------");
    }
    //从键盘上读取一个整数
    public static int inputInt(){
        int i=0;
        Scanner in=null;
        try {
            in=new Scanner(System.in);
            i=in.nextInt();
        } catch(Exception e){
            System.out.println("Exception when read String from the keyboard.");
            inputString();
        }
        return i;
    }
    //从键盘上读取一个实数
    public static double inputDouble(){
        double d=0;
        Scanner in=null;
        try {
            in=new Scanner(System.in);
            d=in.nextDouble();
        } catch(Exception e){
            System.out.println("Exception when read String from the keyboard.");
            inputString();
        }
        return d;
```

```
    }
    //从键盘上读取一个字符串
    public static String inputString(){
        String s=null;
        Scanner in=null;
        try {
            in=new Scanner(System.in);
            s=in.nextLine();
        } catch(Exception e){
            System.out.println("Exception when read String from the keyboard.");
            inputString();
        }
        return s.trim();
    }
}
```

5．主类 MainClass.java 及项目结构

定义主类文件 MainClass.java，完成整个项目，如图 3.16 所示。

图 3.16　项目结构图

参考代码主要部分如下（主方法部分）：

```
//为卡号为 0101 的用户开户
Card card=new Card("0101");
User user=new User("张三",'男',20,card);
Account acount=new Account("1234",1000,user);

//用户登录
System.out.print("请输入卡号:");
String cardNumber=Tools.inputString();
//验证卡号是否正确
while(!cardNumber.equals(acount.getUser().getCard().getCardNumber()))
{
    System.out.print("卡号错误,请重新输入:");
```

```
            cardNumber=Tools.inputString();
        }
        System.out.print("请输入密码:");
        String password=Tools.inputString();

        //验证密码是否正确
        while(!password.equals(acount.getPassword())){
            System.out.print("密码错误,请重新输入:");
            password=Tools.inputString();
        }
        while(true){
            Tools.showATMInfo();
            int i=Tools.inputInt();
            switch(i){
              case 1:
                acount.saveMoney();      break;
              case 2:
                acount.outMoney();       break;
              case 3:
                acount.showMoney();      break;
              case 4:
                 acount.modifyPassword(); break;
              case 5:
                Tools.show(acount);break;
              case 6:
                System.out.println("系统已经退出,再见!");return;
              default:
                 Tools.showATMInfo();
            }
        }
    }
```

3.3.3 项目测试与改进

1. 项目运行与测试

(1) 完成项目文件的编写后,运行项目(Ctrl+F11键),运行结果如图3.17所示。

(2) 分别输入正确的卡号和不正确的卡号测试程序,应该分别进入输入密码流程(见图3.18)和提示"卡号错误,请重新输入:"流程(见图3.19)。

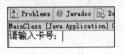

图3.17 输入卡号   图3.18 卡号输入正确   图3.19 卡号输入错误

(3) 在输入密码阶段,分别输入正确的密码和不正确的密码,应该分别进入ATM系统主界面(见图3.20)和提示"密码错误,请重新输入:"流程(见图3.21)。

图 3.20　ATM 系统主界面　　　　图 3.21　密码输入错误

依次选择 1～5 功能并进行测试,总结项目运行结果和设计目标是否一致,并思考项目中存在哪些问题。

2. 项目改进

上述的 ATM 管理系统程序基本实现了项目需求,在此基础上还可以考虑从以下几个方面进行改进。

(1) 为程序增加金额数据输入合法性检查功能:在用户输入存取款金额时,如果输入的数据不是数值型数据,应提示用户数据输入错误,并提示应该输入数值型数据。

(2) 改进程序界面:可以将程序界面改成 GUI 方式的,用户体验会更友好。

3.4　课后训练项目:银行业务调度系统

1. 需求描述

应用 Java 编程语言和面向对象开发方法,编程模拟实现银行业务调度系统逻辑,具体需求如下。

- 银行内有 6 个业务窗口,1～4 号窗口为普通窗口,5 号窗口为快速窗口,6 号窗口为 VIP 窗口。
- 有三种对应类型的客户:VIP 客户、普通客户、快速客户(办理如交水电费、电话费之类业务的客户)。
- 异步随机生成各种类型的客户,生成各类型用户的概率比例为:

　　　　　　　　VIP 客户:普通客户:快速客户=1:6:3
- 客户办理业务所需时间有最大值和最小值,在该范围内随机设定每个 VIP 客户以及普通客户办理业务所需的时间,快速客户办理业务所需时间为最小值(提示:办理业务的过程可通过线程 Sleep 的方式模拟)。
- 各类型客户在其对应窗口按顺序依次办理业务,当 VIP(6 号)窗口和快速业务(5 号)窗口没有客户等待办理业务的时候,这两个窗口可以处理普通客户的业务,而一旦有对应的客户等待办理业务的时候,则优先处理对应客户的业务。
- 随机生成客户时间间隔以及业务办理时间最大值和最小值自定,可以设置。
- 不要求实现 GUI,只考虑系统逻辑实现,可通过 Log 方式展现程序运行结果。

2. 系统分析提示

有三种对应类型的客户：VIP 客户、普通客户、快速客户，异步随机生成各种类型的客户，各类型客户在其对应窗口按顺序依次办理业务。

每一个客户其实就是由银行的一个取号机器产生号码的方式来表示的。所以，可以设计一个号码管理器对象，让这个对象不断地产生号码，就等于随机生成了客户。

由于有三类客户，每类客户的号码编排都是完全独立的，所以，一共要产生三个号码管理器对象，各自管理一类用户的排队号码。这三个号码管理器对象统一由一个号码机器进行管理，这个号码机器在整个系统中始终只能有一个，所以，它要被设计成单例。

各类型客户在其对应窗口按顺序依次办理业务，准确地说，应该是窗口依次叫号，服务窗口每次找号码管理器获取当前要被服务的号码。

根据上面的分析，可以设计以下类实现系统业务逻辑：

1) NumberManager 类

定义一个用于存储上一个客户号码的成员变量和用于存储所有等待服务的客户号码的队列集合。

定义一个产生新号码的方法和获取马上要为之服务的号码的方法，这两个方法被不同的线程操作了相同的数据，所以，要进行同步。

2) NumberMachine 类

定义三个成员变量分别指向三个 NumberManager 对象，分别表示普通、快速和 VIP 客户的号码管理器，定义三个对应的方法来返回这三个 NumberManager 对象。

将 NumberMachine 类设计成单例。

3) CustomerType 枚举类

系统中有三种类型的客户，所以用定义一个枚举类，其中定义三个成员分别表示三种类型的客户。

重写 toString 方法，返回类型的中文名称。这是在后面编码时重构出来的，刚开始不用考虑。

4) ServiceWindow 类

定义一个 start 方法，内部启动一个线程，根据服务窗口的类别分别循环调用三个不同的方法。

定义三个方法分别对三种客户进行服务，为了观察运行效果，应详细打印出其中的细节信息。

5) MainClass 类

用 for 循环创建出四个普通窗口，再创建出一个快速窗口和一个 VIP 窗口。

接着再创建三个定时器，分别定时去创建新的普通客户号码、新的快速客户号码、新的 VIP 客户号码。

6) Constants 类

定义三个常量：MAX_SERVICE_TIME、MIN_SERVICE_TIME、COMMON_CUSTOMER_INTERVAL_TIME。

第 4 章

Java 在线考试系统设计与实现

本章将应用 J2SE 和 MySQL 数据库完成一个 Java 在线考试系统项目的开发。之所以选择考试系统,是因为大家比较熟悉该系统功能,能很容易地理解业务流程,更多地关注设计方法和技术实现,熟悉 Java 数据访问技术和"分层"设计思想,从而掌握一般软件项目的设计思路。

4.1 系统分析

一个软件项目的开发是从分析用户需求开始的。本节将对考试系统软件项目进行简单的需求分析、业务流程分析和数据分析,从而明确项目需求和业务流程,并为数据库设计做好准备。

4.1.1 需求分析

随着计算机和网络的普及应用,传统的卷面考试方式已经不能适应现代考试的需要,也不利于学生随时把握知识掌握情况。建立一个便利的在线考试平台,提供学生在线考试和评阅、教师在线试题维护和管理,以及教师、学生个人信息的管理,实现考试、评分、管理一体化与信息化,可以提高考生学习效率,降低老师评卷工作量,减少管理成本。

具体需求如下:

1. 按用户类别存储个人信息

要求对考生、教师和管理员用户分别存储其个人信息:记录考生用户的考生号、姓名、性别、身份证号、密码、系别和电话号码;记录教师用户的教工号、姓名、性别、身份证号、密码、职称和电话号码;记录管理员用户的管理员号、姓名、性别、身份证号、密码和电话号码。

学生和教师用户可以有多个,其信息由管理员用户创建、删除和修改;而管理员用户在系统运行后不能增加、删除和修改。

2. 按试题类型存储试题信息

要求分别存储填空题、选择题及其答案信息:题号顺序从小到大,不允许空号,从 1

开始;填空题记录题干和答案;选择题记录题干、四个选项和答案序号(A、B、C、D)。

试题可以在系统运行过程中由教师用户维护,如增加、删除和修改;学生用户和管理员用户不能修改试题。

3. 按用户类别登录进入系统

从同一界面登录到系统,进行用户身份的验证并根据用户身份进入不同的操作界面:学生用户进入考试界面,可以进行答题和修改个人密码;教师用户进入教师管理界面,可以进行试题维护和修改个人密码;管理员用户进入系统管理界面,可以进行学生和教师用户信息的增删改和修改个人密码。

4. 试题随机生成并自动判分

由系统在试题库中随机抽取题目组成由填空题和选择题构成的试卷,考试结束由系统自动判分并反馈给考生。

5. 实现考试计时

考生进入考试系统则开始计时并显示给考生,允许考生在允许的时间范围内主动交卷或在达到考试时间后由系统强行收卷;考试时间剩余5分钟的时候给学生提示。

6. 统一而有效的数据存储

实现系统数据的永久存储,试题数据、用户信息等全部统一存储在安全的数据库服务器系统中,所有用户访问的是相同的数据;存取数据迅速有效、代价低。

4.1.2 业务流程分析

为实现系统要求的功能,经分析应具备的相关业务主要包括学生在线考试及评分、教师在线维护试题、管理员管理用户。业务流程图如图4.1所示。

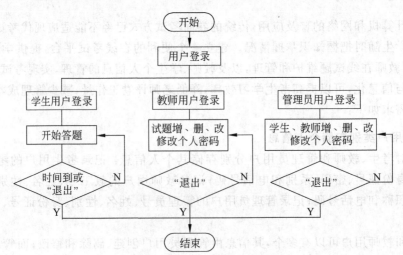

图 4.1 考试系统业务流程图

4.1.3 数据分析

由于考试系统项目比较简单,经过前面的分析,涉及的数据主要包括两大类别:用户信息和试题信息。这两类数据在系统之间没有复杂的关联,只需按照需求设计相应的数据实体即可。

4.2 系统设计

4.2.1 系统设计思路

本系统采用了分层化模块设计,以系统公用数据库为操作对象,将整个系统划分为考生管理模块、教师管理模块、系统管理模块。

由于本项目规模小,数据量也不大,业务逻辑较简单,设计用 MySQL 数据库实现数据的永久存储;将不同的用户、试题等数据以及对这些数据的操作封装到不同的类中(实体类);将对数据库和数据表的操作也封装起来(数据访问类);再定义用户操作界面(视图类)调用这些方法实现相应模块的功能。

将系统设计为分层实现,可以使系统实现起来更加容易;另外,也更容易改进系统,例如,可以很容易将这个系统改造为 B/S 架构的 Web 应用程序,只需要将用户界面部分改用某种网页形式(如 JSP)就可以了,而且分层结构也是复杂软件项目基本的设计思路。

4.2.2 功能模块设计

依据前面的设计思路,把在线考试系统划分为三大模块,分别为考生管理模块、教师管理模块、系统管理模块。系统功能结构图如图 4.2 所示。

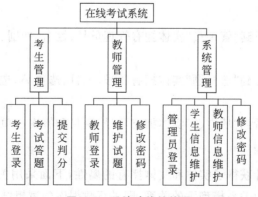

图 4.2 系统功能结构图

4.2.3 数据库设计

1. 概念结构设计

依据前面的数据分析和需求，抽象出考生、教师和管理员用户实体和填空题、选择题试卷信息实体共六个，它们的 E-R 图如图 4.3～图 4.7 所示。

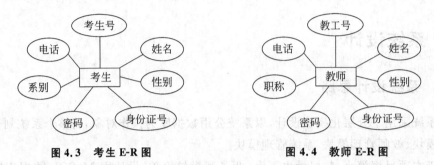

图 4.3　考生 E-R 图　　　　　　　图 4.4　教师 E-R 图

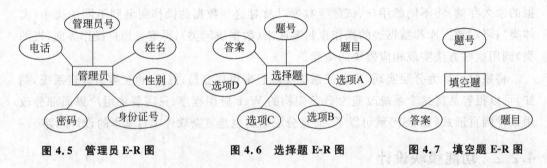

图 4.5　管理员 E-R 图　　　图 4.6　选择题 E-R 图　　　图 4.7　填空题 E-R 图

从图 4.3 中可以看到"考生"实体拥有考生号、姓名、性别、身份证号、密码、系别和电话共七项属性。

从图 4.4 中可以看到"教师"实体拥有教工号、姓名、性别、身份证号、密码、职称和电话七项属性。

从图 4.5 中可以看到"管理员"实体拥有管理员号、姓名、性别、身份证号、密码和电话六项属性。

从图 4.6 中可以看到"选择题"实体拥有题号、题目、选项 A、选项 B、选项 C、选项 D、和答案七项属性。

从图 4.7 中可以看到"填空题"实体拥有题号、题目和答案三项属性。

2. 逻辑结构设计

通过概念结构设计获得了实体和实体拥有的属性，下面采用关系型数据库把概念设计模型中的实体转换为关系模型下的关系，即进行数据库的逻辑结构设计。

逻辑结构设计得出的五种实体关系如表 4.1 所示。

表 4.1 逻辑结构设计实体关系表

关系表名	用途
Student	学生实体关系表,用于保存本系统的考生信息
Teacher	教师实体关系表,用于保存本系统的教师信息
Administrator	管理员实体关系表,用于保存本系统的管理员信息
FillQuestion	填空题实体关系表,用于保存考试系统题库中的填空题及答案
ChoiceQuestion	选择题实体关系表,用于保存考试系统题库中的选择题及答案

1) 学生实体关系表(Student)说明(见表 4.2)

表 4.2 学生实体关系表

字段名	类型	长度	可为空	约束条件	注释
sid	varchar	6	否	PK	主键、考生登录 ID
sname	varchar	8	否		考生姓名
sex	varchar	2			考生性别
cardnumber	varchar	18			身份证号
pwd	varchar	3	否		考生密码
department	varchar	20			系别
phone	varchar	11			考生电话

2) 教师实体关系表(Teacher)说明(见表 4.3)

表 4.3 教师实体关系表

字段名	类型	长度	可为空	约束条件	注释
tid	varchar	6	否	PK	主键、教师登录 ID
tname	varchar	8	否		教师姓名
sex	Varchar	2			教师性别
cardnumber	varchar	18			身份证号
pwd	varchar	6	否		教师用户密码
title	varchar	4			职称
phone	varchar	11			电话

3) 管理员实体关系表（Administrator）说明（见表4.4）

表4.4 教师管理员实体关系表

字段名	类型	长度	可为空	约束条件	注释
adid	varchar	6	否	PK	主键、管理员登录ID
aname	varchar	8	否		管理员姓名
sex	Varchar	2			管理员性别
cardnumber	varchar	18			身份证号
pwd	varchar	8	否		管理员用户密码
phone	varchar	11			电话

4) 填空题实体关系表（FillQuestion）说明（见表4.5）

表4.5 填空题实体关系表

字段名	类型	长度	可为空	约束条件	注释
f_id	int		否	PK	题目编号、主键、自增
f_question	varchar	200	否		题目、唯一
f_answer	varchar	50			答案

5) 选择题实体关系表（ChoiceQuestion）说明（见表4.6）

表4.6 选择题实体关系表

字段名	类型	长度	可为空	约束条件	注释
c_id	int		否	PK	题目编号、主键、自增
c_question	varchar	200	否		题目、唯一
c_choiceA	varchar	100			选项A
c_choiceB	varchar	100			选项B
c_choiceC	varchar	100			选项C
c_choiceD	varchar	100			选项D
c_answer	varchar	2			答案

4.2.4 类的分层设计

依据前面的设计思路，在线考试系统采用分层设计模式，可设计实体类（entity）、数据访问类（dao）和视图类（view）三种不同性质的类，并将其分别定义在entity、dao和gui包中，使逻辑层次更清晰。

1. 实体类

实体类也称Bean类，因通常代表数据表中的数据实体信息而得名，实体类一般包含与数据表结构相同的私有成员变量和对这些变量进行操作的公有get和set方法。

本项目中的实体类用来封装用户和试题信息变量和对这些变量的操作方法,具体如下:
用户信息类 Student、Teacher、Administrator 如图 4.8 所示。

```
        Student                              Teacher
-sid : String                        -tid : String
-name : String                       -name : String
-sex : String                        -sex : String
-cardNumber : String                 -cardNumber : String
-password : String                   -password : String
-dept : String                       -title : String
-phone : String                      -phone : String
+get、set方法() : <未指定>            +get、set方法() : <未指定>
   (a) Student考生实体类                 (b) Teacher教师实体类

             Administrator
      -adid : String
      -name : String
      -sex : String
      -cardNumber : String
      -password : String
      -phone : String
      +get、set方法() : <未指定>
              (c) 管理员实体类
```

图 4.8　用户信息类 Student、Teacher、Administrator

试题信息类 FillQuestion、ChoiceQuestion 如图 4.9 所示。

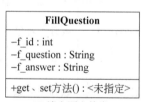
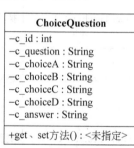

(a) 填空题实体类　　　(b) 选择题实体类

图 4.9　试题信息类 FillQuestion、ChoiceQuestion

2. 数据访问类

数据访问类因操作数据库中的数据表而得名,主要封装了需要进行数据持久化操作的各个方法,通常与数据表及实体类一一对应。本项目中的数据访问类用来封装对用户和试题数据表进行数据访问操作的方法,各方法的具体说明如表 4.7～表 4.12 所示。

1) 数据库连接工厂类 JDBCConnectionFactory

表 4.7　数据库连接工厂类方法说明

方　法	功　能	参数说明	返回值说明
public static Connection getConnection()	建立数据库连接	无	数据库连接对象

2) 考生数据访问类 StudentDAO

表 4.8　StudentDAO 类方法说明

方法	功能	参数说明	返回值说明
public Student selectStudent(String sid)	根据学号查询学生	学号	学生类对象
public int insertStudent(Student st)	插入一条学生记录	学生类对象	1：学生已存在 0：插入成功 —1：插入不成功
public int deleteStudent(Student st)	删除一条学生记录	学生类对象	1：学生不存在 0：删除成功 —1：删除不成功
public int updateStudent(Student st)	更新学生记录	学生类对象	1：学生不存在 0：更新成功 —1：更新不成功

3) 教师数据访问类 TeacherDAO

表 4.9　TeacherDAO 类方法说明

方法	功能	参数说明	返回值说明
public Teacher selectTeacher(String tid)	根据教工号查询教师	教工号	教师类对象
public int insertTeacher(Teacher te)	插入一条教师记录	教师类对象	1：教师已存在 0：插入成功 —1：插入不成功
public int deleteTeacher(Teacher st)	删除一条教师记录	教师类对象	1：教师不存在 0：删除成功 —1：删除不成功
public int updateTeacher(Teacher te)	更新教师记录	教师类对象	1：教师不存在 0：更新成功 —1：更新不成功

4) 管理员数据访问类 AdministratorDAO

表 4.10　AdministratorDAO 类方法说明

方法	功能	参数说明	返回值说明
public Administrator electAdministrator(String adid)	根据管理员号查询	管理员号	管理员类对象
public int updateAdministrator(Administrator ad)	更新管理员记录	管理员类对象	1：管理员不存在 0：更新成功 —1：更新不成功

5）填空题数据访问类 FillQuestionDAO

表 4.11　FillQuestionDAO 类方法说明

方　　法	功　　能	参 数 说 明	返 回 值 说 明
public List＜FillQuestion＞selectFillQuestion（int num）	随机选择填空题	题目数量	保存了试题的链表
public int insertFillQuestion（FillQuestion fill）	插入一个填空题	填空题类对象	1：试题已存在 0：插入成功 －1：插入不成功
public int deleteFillQuestion（FillQuestion fill）	删除一条填空题记录	填空题类对象	1：填空题不存在 0：删除成功 －1：删除不成功
public int updateFillQuestion（FillQuestion fill）	更新填空题记录	填空题类对象	1：填空题不存在 0：更新成功 －1：更新不成功

6）选择题数据访问类 ChoiceQuestionDAO

表 4.12　ChoiceQuestionDAO 类方法说明

方　　法	功　　能	参 数 说 明	返 回 值 说 明
public List＜ChoiceQuestion＞ selectChoiceQuestion（int num）	随机选择选择题	题目数量	保存了试题的链表
public int insertChoiceQuestion（ChoiceQuestion choice）	插入一个选择题	选择题类对象	1：试题已存在 0：插入成功 －1：插入不成功
public int deleteChoiceQuestion（ChoiceQuestion choice）	删除一条选择题记录	选择题类对象	1：选择题不存在 0：删除成功 －1：删除不成功
public int updateChoiceQuestion（ChoiceQuestion choice）	更新选择题记录	选择题类对象	1：选择题不存在 0：更新成功 －1：更新不成功

3．图形用户界面视图类

实现系统运行的功能界面视图，可以用不同的方法实现，如网页、Java 程序类等。这是采用了 Java 类实现用户界面，在这些视图类中处理用户对界面的操作事件，调用数据访问类方法，完成数据的处理，进而完成系统功能。

根据功能模块设计及业务流程，设计图形用户界面视图类如表 4.13～4.18 所示。

1) 用户登录界面视图类 LoginFrame

表 4.13 LoginFrame 类设计

功能	实现考生、教师、管理员用户登录功能
界面要素	"关于"菜单 欢迎语标签；用户类型、用户 ID、用户口令标签 选择用户类型下拉列表；用户 ID 文本输入框、密码输入框 登录、清屏、退出按钮
设计要点	布局合理，窗体大小固定 菜单项动作事件显示 About 窗体视图，介绍开发者信息 登录按钮动作事件处理登录验证 清屏按钮动作事件清除用户输入文本框的内容 退出按钮动作事件退出系统

2) "关于"界面视图类 AboutFrame

表 4.14 AboutFrame 类设计

功能	实现显示系统开发的作者、地址、联系方式等信息功能
界面要素	显示作者、地址、联系方式信息
设计要点	布局合理，窗体大小固定 分行显示

3) 开始考试界面视图类 BeginTest

表 4.15 BeginTest 类设计

功能	实现考生登录后显示个人信息并开始计时考试功能
界面要素	"欢迎登录"提示标签；"当前用户姓名"、"身份证号"标签 显示当前用户的名字和身份证号的组件(值不可修改) "开始考试并计时"、"取消登录"按钮
设计要点	布局合理，窗体大小固定 "开始考试并计时"按钮动作事件应提示考试时间、考题数量、分值等信息，然后显示"考生管理界面"窗体并使当前窗体不可见 "取消登录"按钮动作事件返回"用户登录界面"

4) 考生管理界面视图类 StudentFrame

表 4.16 StudetFrame 类设计

功能	实现考生在线答题功能
界面要素	实时显示考试时间的标签，时间要突出显示 "交卷"按钮 "选择题"和"填空题"在不同的选项卡显示，选项卡中有试题区和答题区以及上下切换试题的按钮 "选择题"选项卡中有显示选择题答题记录的文本区

设计要点	布局合理,窗体大小固定
	计时功能可通过多线程实现,以秒计时 倒计时剩余 5 分钟时,显示提示用户时间的消息对话框 时间剩余 0 时,计算总成绩并显示在消息对话框中
	"交卷"按钮动作事件处理计算总成绩并显示在消息对话框中 上下切换题目的按钮动作事件处理试题的切换显示,应判断是否第一题和最后一题

5) 教师管理界面视图类 TeacherFrame

表 4.17　TeacherFrame 类设计

功能	实现教师在线管理试题、修改个人密码功能
界面要素	"选择题"、"填空题"、"个人信息修改"在不同的选项卡显示 "选择题"选项卡提供题目编号、内容、四个选项、答案的输入文本框和提示标签,还有增、删、改、查及退出按钮 "填空题"选项卡提供题目编号、内容、答案的输入文本框和提示标签,还有增、删、改、查及退出按钮 "个人信息"选项卡提供原口令、新口令、确认口令文本框和提示标签,还有确认、清屏及退出按钮
设计要点	布局合理,窗体大小固定 修改密码实现合法性判定:原口令要正确,新口令不能为空,新口令和确认口令要一致 "增、删、改、查"按钮动作事件处理需正确调用试题数据访问类方法 "退出"按钮动作事件,应使系统从任何位置退出

6) 系统管理界面视图类 AdminFrame

表 4.18　AdminFrame 类设计

功能	实现管理员对学生、教师用户的管理和修改个人密码功能
界面要素	"学生管理"、"教师管理"、"个人信息"在不同的选项卡显示 "学生管理"选项卡提供学生 ID、姓名、性别、身份证号、口令、系别及电话的输入文本框和提示标签,还有增、删、改、查及退出按钮 "教师管理"选项卡提供教工 ID、姓名、性别、身份证号、口令、职称及电话的输入文本框和提示标签,还有增、删、改、查及退出按钮 "个人信息"选项卡提供原口令、新口令、确认口令文本框和提示标签,还有确认、清屏及退出按钮
设计要点	布局合理,窗体大小固定 修改密码实现合法性判定:原口令要正确,新口令不能为空,新口令和确认口令要一致 "增、删、改、查"按钮动作事件处理需正确调用试题数据访问类方法 "退出"按钮动作事件,应使系统从任何位置退出

4.3 系统实现与测试

4.3.1 数据库的建立与连接

MySQL 作为一款免费软件，使用非常简单；相对于 Oracle 等数据库来说，在低硬件环境下，MySQL 是比较经济的。这里考试系统采用 MySQL 建立数据库。

下面从数据库环境的构建开始逐步介绍本系统数据库的建立过程。

1. 安装 MySQL 和 MySQL Front 工具

1) 安装 MySQL

(1) 运行 mysql-5.5.21-win32.msi，启动安装向导，如图 4.10 所示。

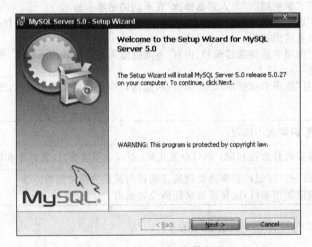

图 4.10　MySQL 安装向导

(2) 选择典型安装模式，如图 4.11 所示。

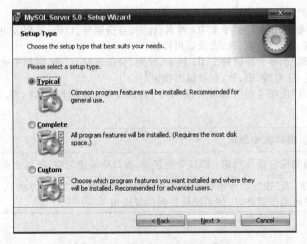

图 4.11　MySQL 典型安装模式

(3) 跳过注册,如图 4.12 所示。

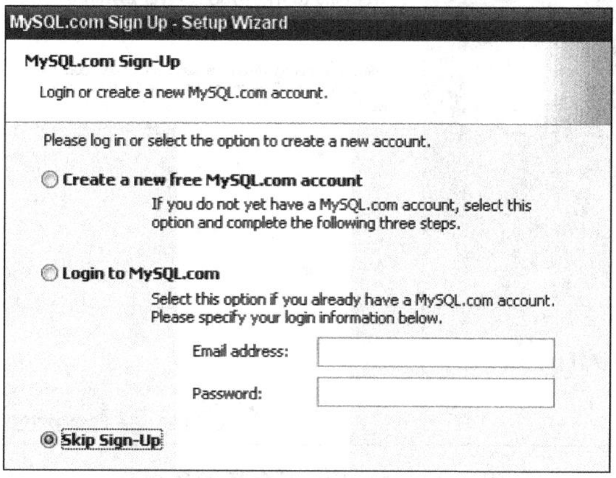

图 4.12　跳过注册

(4) 结束安装,如图 4.13 所示。

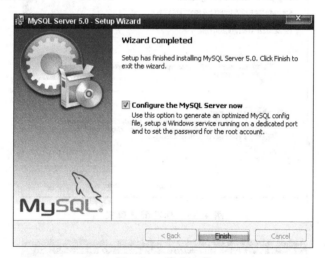

图 4.13　结束安装

单击 Finish 按钮可以结束 MySQL 的安装,如果选择 Configure the MySQL Server Now 复选框,则在安装完成后会紧接着出现 MySQL 服务器的配置界面,让用户完成对 MySQL 数据库服务器的配置。

2) 配置 MySQL 数据库服务器

(1) 进入 MySQL 服务器配置向导,如图 4.14 所示。

(2) 选择标准配置模式,如图 4.15 所示。

(3) 设置 Windows 选项,如图 4.16 所示。

建议将界面中的两个复选框都选中,一个是添加到 Windows 服务中,另一个是包含命令行方式。

(4) 设置安全选项,如图 4.17 所示。

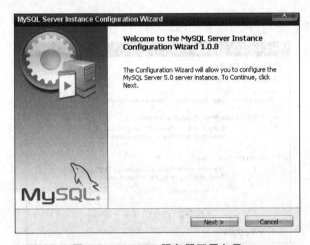

图 4.14 MySQL 服务器配置向导

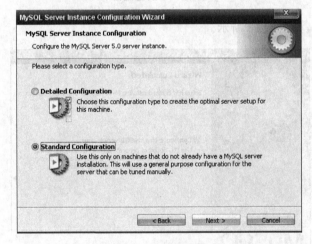

图 4.15 标准配置模式

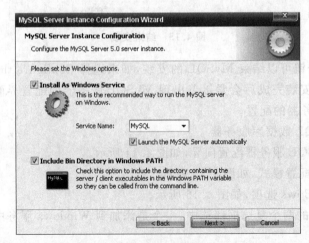

图 4.16 设置 Windows 选项

此步骤中要求为 MySQL 的默认用户 root 设置密码,也可以在此选择创建一个匿名用户。

图 4.17 设置安全选项

(5) 单击 Next 按钮,开始执行配置过程。

3) 安装 MySQL Front

(1) 运行 MySQL-Front_Setup.exe 开始安装,如图 4.18 所示。

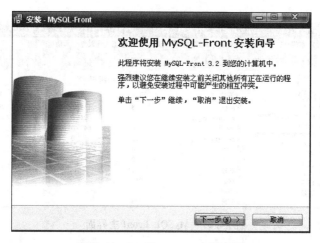

图 4.18 MySQL-Front 安装向导

(2) 单击"下一步"按钮确定安装目录位置,并开始安装即可。

4) 启动 MySQL Front

MySQL Front 是一个可以对 MySQL 进行可视化操作的前台环境,可以在此即学即用的界面中方便地进行各种数据库管理工作。

2. 设置注册信息

从"程序"→MySQL Front→MySQL Front 中运行安装好的 MySQL Front,选取"注

册"选项卡,输入安装时设置的密码,如 yang,用户为默认的 root,如图 4.19 所示。

1) 设置连接

单击"连接"选项卡,设置服务器是 localhost,即使用本机作为数据库服务器,端口号 3306,如图 4.20 所示。

图 4.19　MySQL Front 配置注册　　　　图 4.20　MySQL Front 配置连接

2) MySQL Front 主界面

在 MySQL Front 配置连接界面单击"确定"按钮,即可进入 MySQL Front 主界面,如图 4.21 所示。

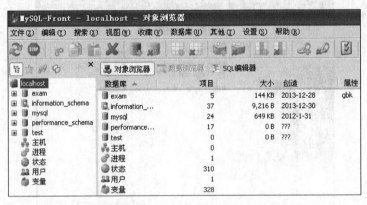

图 4.21　MySQL Front 主界面

3. 创建数据库及数据表

在项目开发中,通常可以使用数据库创建模板文件完成数据库的快速创建,模板文件请参阅 myexam.sql。

本项目中,依据系统设计阶段的数据表设计结果,编辑模板 SQL 文件,可以很快就完成数据库的创建及原始数据的录入,如图 4.22 所示。

单击绿色三角工具按钮 ▶ ,执行该 SQL 文本,即可完成数据库及数据表的创建和初始数据的录入,简单方便。完成后的数据库系统结构如图 4.23 所示。

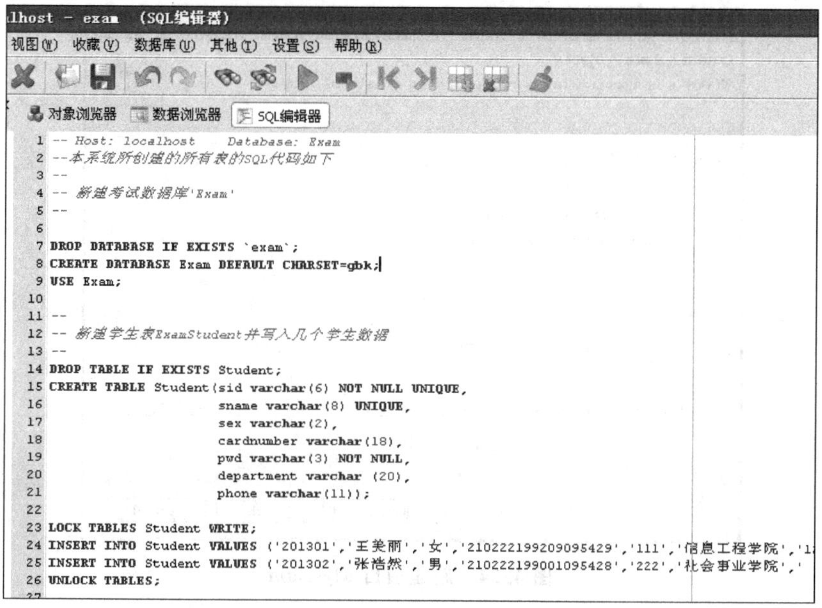

图 4.22　MySQL Front SQL 编辑器

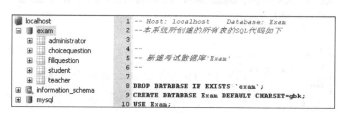

图 4.23　exam 数据库结构

4.3.2　Entity 实体类的实现

面向对象程序设计中,实体类通常与数据表表达的数据实体是一一对应的,当操作数据库中的数据时,在程序中用实体类对象表示这些数据,可以更容易表达。

从本部分开始,将逐步实现项目的创建、包的定义和层次类的定义和测试。

1．新建项目

在 MyEclipse 环境下新建 Java 项目 MyExam,项目属性默认即可,如图 4.24 所示。

2．新建实体类所在包

在项目文件夹下右击 src,选择 New→Package 新建包 entity.exam,如图 4.25 所示。

3．新建实体类

在 entity.exam 包中新建实体类 Student、Teacher、Administrator 以及 FillQuestion、ChoiceQuestion,考虑到三个用户实体类 Student、Teacher、Administrator 具有较多的共同属性,可以定义一个用户超类 User 封装共同的属性,让 Student、Teacher、

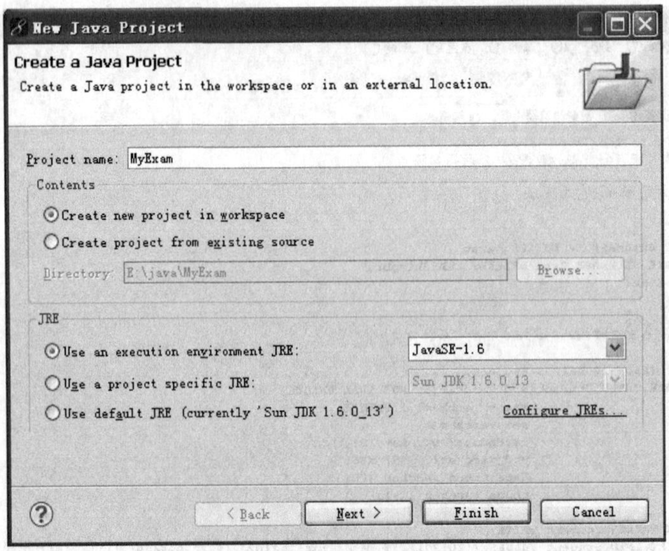

图 4.24 新建项目 MyExam

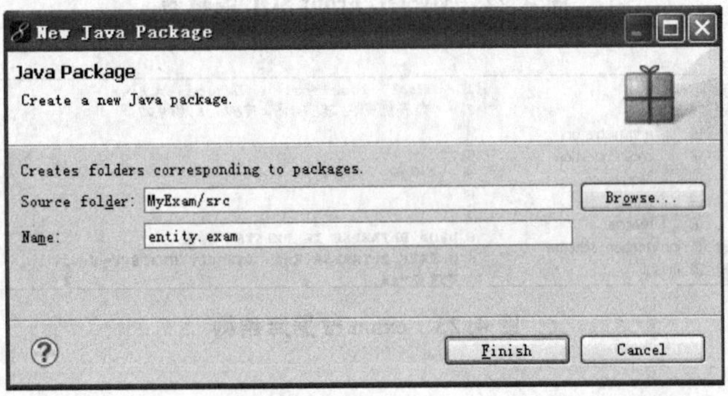

图 4.25 新建包 entity.exam

Administrator 继承 User 类,这样可以实现代码的复用,更能体现面向对象编程的优越性。

这部分的实体类主要封装了对象实体的相关属性数据,并通过公有 get、set 方法操作这些私有属性数据,所以,每个实体类的定义都是相似的。

1) User.java 文件

```
package entity.exam;

//学生、教师、管理员类的父类,能更好地实现代码复用
public class User {
    //这些共同的属性被定义为 protected,可以被子类继承
    protected String name;
    protected String sex;
    protected String cardNumber;
```

```java
    protected String password;
    protected String phone;

    public User(String cardNumber, String name, String sex, String password,
            String phone){
        super();
        this.cardNumber=cardNumber;
        this.name=name;
        this.sex=sex;
        this.password=password;
        this.phone=phone;
    }
    public User(){
        super();
    }
    public String getPhone(){
        return phone;
    }
    public void setPhone(String phone){
        this.phone=phone;
    }
    public String getCardNumber(){
        return cardNumber;
    }
    public void setCardNumber(String cardNumber){
        this.cardNumber=cardNumber;
    }
    public String getName(){
        return name;
    }
    public void setName(String name){
        this.name=name;
    }
    public String getSex(){
        return sex;
    }
    public void setSex(String sex){
        this.sex=sex;
    }
    public String getPassword(){
        return password;
    }
    public void setPassword(String password){
        this.password=password;
```

			}
		}

		2) Student.java 文件

		package entity.exam;

		public class Student extends Person{
			private String sid;
			private String dept;

			public Student(){
				super();
			}
			public Student(String sid, String name, String sex, String cardNumber,
					String password, String dept, String phone){
				super(cardNumber, name, sex, password, phone);
				this.sid=sid;
				this.dept=dept;
			}
			public String getSid(){
				return sid;
			}
			public void setSid(String sid){
				this.sid=sid;
			}
			//其他 get、set 方法略
		}

		3) Teacher.java 文件

		package entity.exam;

		public class Teacher extends Person{
			private String tid;
			private String title;

			public Teacher(String tid,String name, String sex, String cardNumber,
					String password,String title,String phone){
				super(cardNumber, name, sex, password, phone);
				this.tid=tid;
				this.title=title;
			}
			//get、set 方法略
		}

4) Administrator.java 文件

```java
package entity.exam;

public class Administrator extends Person{
    private String adid;

    public Administrator(String adid, String name, String sex,
            String cardNumber, String password, String phone){
        super(cardNumber, name, sex, password, phone);
        this.adid=adid;
    }
    //get、set 方法略
}
```

5) FillQuestion.java 文件

```java
package entity.exam;

public class FillQuestion {
    private int f_id;                      //填空题题号
    private String f_question;             //填空题题目
    private String f_answer;               //填空题答案

    //构造方法
    public FillQuestion(String fQuestion, String fAnswer){
        f_question=fQuestion;
        f_answer=fAnswer;
    }
    public FillQuestion(int fId, String fQuestion, String fAnswer){
        f_id=fId;
        f_question=fQuestion;
        f_answer=fAnswer;
    }
    //get、set 方法略
}
```

6) ChoiceQuestion.java 文件

```java
package entity.exam;

public class ChoiceQuestion {
    private int c_id;                      //选择题号
    private String c_question;             //选择题题干
    private String c_choiceA;              //选项
    private String c_choiceB;
```

```
                private String c_choiceC;
                private String c_choiceD;
                private String c_answer;                    //正确答案

                //构造方法略
                //get、set 方法略
}
```

4.3.3 DAO 数据访问类的实现

DAO 数据访问类主要实现对对象实体数据表的访问,定义相应的数据访问方法实现考试系统需要数据访问功能。在大的复杂系统中,有时,会将数据访问方法和业务逻辑方法分别在 Dao 层和 Service 层实现。

因为考试系统逻辑简单,项目规模也不大,所以没有进一步划分,而只是定义了 DAO 层实现简单的数据访问方法,部分业务逻辑分别在 Dao 类和视图层类中实现。

1. 新建 DAO 类所在包

在项目文件夹下新建包 dao.exam,项目结构如图 4.26 所示。

2. 添加 MySQL JDBC 驱动到构建路径

项目需要引用 JDBC 驱动程序文件 mysql-connector-java-5.0.6-bin.jar,获取该文件后请在项目文件夹下新建 lib 文件夹存放此文件,而不要将此文件放在项目文件夹之外的位置,防止项目移动后引用的文件找不到,如图 4.27 所示。

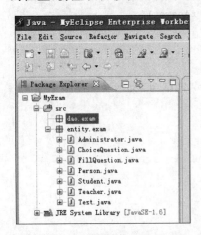

图 4.26 新建包 dao.exam

图 4.27 JDBC 驱动程序文件

MyEclipse 中刷新项目后将会在项目结构目录中看到此目录和文件,如图 4.28 所示。

右击项目 MyExam,选择 Build Path→Configure Build Path,打开项目属性窗口配置构建路径(Java Build Path),如图 4.29 所示。

选择 Add JARs…添加项目目录下 lib 文件中的 jar 文件,如图 4.30 所示,单击 OK

按钮,这样在项目中就可以使用 MySQL JDBC 驱动程序了。

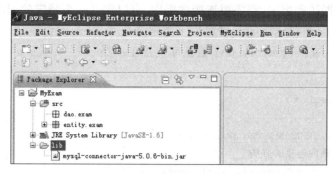

图 4.28 项目文件夹下的 jar 文件

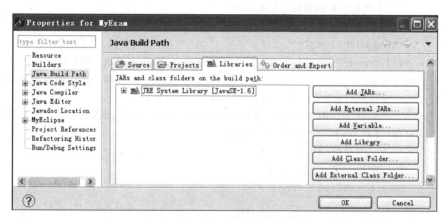

图 4.29 Java Build Path 窗口

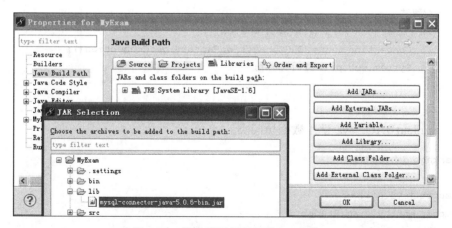

图 4.30 添加项目目录下的 jar 文件

3. 新建 DAO 数据访问类

DAO 数据访问类通常与实体类是一一对应的,此类用来封装对数据实体的数据库访问操作,可以根据项目需要定义各数据访问方法,主要是查询、添加、删除、修改方法,程序的结构和实现也是相似的。

在所有的数据访问类中，都需要连接数据库，可以定义一个特定的JDBC数据库连接工厂类JDBCConnectionFactory。请将以下类定义在dao.exam包中。

1) JDBCConnectionFactory.java 数据库连接工厂类

```java
package dao.exam;
import java.sql.*;

public class JDBCConnectionFactory {
    //获取数据库连接的方法
    public static Connection getConnection(){
        Connection conn=null;
        //连接MySQL数据库的字符串URL
        String url="jdbc:mysql://localhost:3306/Exam";
        String usr="root";                          //MySQL用户名
        String pwd="yang";                          //MySQL用户口令
        try {                                       //加载MySQL数据库驱动程序
            Class.forName("com.mysql.jdbc.Driver");
            conn=DriverManager.getConnection(url, usr, pwd);   //建立连接
        }catch(ClassNotFoundException e){
            e.printStackTrace();
        }catch(SQLException e){
            e.printStackTrace();
        }
        return conn;
    }
    //测试
    public static void main(String args[]){
        Connection conn=JDBCConnectionFactory.getConnection();
        System.out.println(conn);
    }
}
```

2) StudentDAO.java 学生数据访问类

```java
package dao.exam;

import java.sql.*;
import entity.exam.Student;

public class StudentDAO {
    //根据学号查询学生信息
    public Student selectStudent(String sid){
        Connection conn=JDBCConnectionFactory.getConnection();
        Student st=null;
```

```java
        //根据学号查询学生
        try {
            Statement stmt=conn.createStatement();
            String sql="select *  from student where  sid='"+sid+"'";
            ResultSet rs=stmt.executeQuery(sql);
            if(rs.next()){
                st=new Student(rs.getString(1),rs.getString(2),rs.getString
                    (3),rs.getString(4),rs.getString(5),rs.getString(6),rs.
                    getString(7));
            }
        } catch(SQLException ee){
            ee.printStackTrace();
        } finally {                                    //关闭连接
            if(conn !=null){
                try {
                    conn.close();
                } catch(SQLException e){
                    e.printStackTrace();
                }
            }
        }
        return st;
    }
    /*
     * 插入一条学生记录；
     * 返回值为1该学生已存在；
     * 返回值为0表示插入成功；
     * 返回值为-1表示插入数据不成功
     */
    public int insertStudent(Student st){
        if(selectStudent(st.getSid())!=null)           //如学生记录已存在
            return 1;
        Connection conn=JDBCConnectionFactory.getConnection();
        try {
            //插入一条记录
            String str="insert into student values"+"('"+st.getSid()
                +"','"+st.getName()+"','"+st.getSex()+"','"
                +st.getCardNumber()+"','"+st.getPassword()+"','"
                +st.getDept()+"','"+st.getPhone()+"')";
            PreparedStatement pstmt=conn.prepareStatement(str);
            if(pstmt.executeUpdate()==1){
                return 0;
            } else {
                return-1;
```

```
            }
        } catch(SQLException ee){
            return-1;
        } finally {                                    //关闭连接
            if(conn !=null){
                try {
                    conn.close();
                } catch(SQLException e){
                    e.printStackTrace();
                }
            }
        }
    }

    /*
     * 更新学生记录；
     * 返回值为1该学生不存在；
     * 返回值为0表示更新成功；
     * 返回值为-1表示更新数据不成功
     */
    public int updateStudent(Student st){
        if(selectStudent(st.getSid())==null)        //要更新的学生不存在
            return 1;
        Connection conn=JDBCConnectionFactory.getConnection();
        try {
            //更新记录
            String str="update student set sname='"+st.getName()
                +"',sex='"+st.getSex()+"',cardnumber='"
                +st.getCardNumber()+"',pwd='"+st.getPassword()
                +"',department='"+st.getDept()+"',phone='"
                +st.getPhone()+"' where sid='"+st.getSid()+"'";
            PreparedStatement pstmt=conn.prepareStatement(str);
            if(pstmt.executeUpdate()==1){
                return 0;
            } else {
                return-1;
            }
        } catch(SQLException ee){
            ee.printStackTrace();
            return-1;
        } finally {                                    //关闭连接
            if(conn !=null){
                try {
                    conn.close();
```

```java
            } catch(SQLException e){
                e.printStackTrace();
            }
        }
    }
}
/*
 * 删除学生记录；
 * 返回值为1 该学生不存在；
 * 返回值为0 表示删除成功；
 * 返回值为-1 表示删除记录不成功
 */
public int deleteStudent(Student st){
    if(selectStudent(st.getSid())==null)          //要删除的学生不存在
        return 1;
    Connection conn=JDBCConnectionFactory.getConnection();
    try {
        //删除记录
        String str="delete from student where sid='"+st.getSid()+"'";
        PreparedStatement pstmt=conn.prepareStatement(str);
        if(pstmt.executeUpdate()==1){
            return 0;
        } else {
            return-1;
        }
    } catch(SQLException ee){
        ee.printStackTrace();
        return-1;
    } finally {                                    //关闭连接
        if(conn !=null){
            try {
                conn.close();
            } catch(SQLException e){
                e.printStackTrace();
            }
        }
    }
}

//测试这个类,项目中可以没有
public static void main(String args[]){
    Student st= new Student("201302","陈文","男","2111111199009091234",
        "111","信息工程学院","13222222334");
    StudentDAO dao=new StudentDAO();
```

```
            int i=dao.deleteStudent(st);
            System.out.println(i);
        }
    }
```

3) TeacherDAO 教师数据访问类

```
package dao.exam;

import java.sql.*;
import entity.exam.Teacher;

public class TeacherDAO {
    //根据教师号查询教师信息
    public Teacher selectTeacher(String tid){
        //与学生数据访问类中的 selectStudent 方法类似,请参照完成
    }
    //插入一条教师记录
    public int insertTeacher(Teacher te){
        //与学生数据访问类中的 insertStudent 方法类似,请参照完成
    }
    //更新教师记录
    public int updateTeacher(Teacher te){
        //与学生数据访问类中的 updateStudent 方法类似,请参照完成
    }
    //删除教师记录
    public int deleteTeacher(Teacher st){
        //与学生数据访问类中的 deleteStudent 方法类似,请参照完成
    }
}
```

4) AdministratorDAO 管理员数据访问类

```
package dao.exam;

import java.sql.*;
import entity.exam.Administrator;

public class AdministratorDAO {
    //根据管理员号查询管理员信息
    public Administrator selectAdministrator(String adid){
        //与学生数据访问类中的 selectStudent 方法类似,请参照完成
    }
    //更新管理员信息
    public int updateAdministrator(Administrator ad){
        //与学生数据访问类中的 updateStudent 方法类似,略
```

		}
}

5）ChoiceQuestionDAO 选择题数据访问类

```
package dao.exam;

import java.sql.*;
import java.util.Iterator;
import java.util.LinkedList;
import java.util.List;
import entity.exam.ChoiceQuestion;

public class ChoiceQuestionDAO {
	/*
	 * 这个方法可以产生多个不同的随机正整数,作为从题库中抽取试题的题目编号
	 * 参数:max 表示随机数的最大值,在本项目中代表试题库中题目数;
	 *      num 表示产生的随机正整数个数;
	 * 返回值:存储了 num 个随机正整数的整型数组
	 */
	private int[] random_number(int max, int num){
		int[] number=new int[num];
		for(int i=0; i<num; i++){
			number[i]=(int)(Math.random() * max)+1;
			//如果当前产生的随机数与已经产生的随机数相同,则重新生成
			for(int j=1; j<=i; j++){
				if(number[j-1]==number[i]){
					i=i-1;    break;
				}
			}
		}
		return number;
	}

	//从题库中生成随机试题,返回保存 num 道选择题的链表
	public List<ChoiceQuestion> selectChoiceQuestion(int num){
		Connection conn=JDBCConnectionFactory.getConnection();
		ChoiceQuestion choice=null;
		List<ChoiceQuestion> list=new LinkedList<ChoiceQuestion>();

		//查询该选择题
		try {
			Statement stmt=conn.createStatement();
			//生成 num 个随机数作为选题的题号
			ResultSet rs1= stmt.executeQuery(" select max(c _ id) from " +
```

```java
            "choiceQuestion");
        int maxcid=30;                                    //默认最大题目编号为 30
        if(rs1.next()){
            maxcid=rs1.getInt(1);                         //根据查询结果设置最大题目编号
        }
        int[] number=random_number(maxcid, num);          //生成选题编号数组

        //查询所有选择题并选出 num 道题
        String sql="select *  from choiceQuestion";
        ResultSet rs=stmt.executeQuery(sql);
        while(rs.next()){
            for(int i=0; i<num; i++){
                if(rs.getInt(1)==number[i]){
                    choice=new ChoiceQuestion(rs.getString(2),rs.getString
                        (3),rs.getString(4),rs.getString(5),rs.getString(6),
                        rs.getString(7));
                    list.add(choice);
                }
            }
        }
    } catch(SQLException ee){
        ee.printStackTrace();
    } finally {
        //关闭连接,略
    }
    return list;
}
//根据题号查询一道选择题
public ChoiceQuestion selectOneChoiceQuestion(int cid){
    Connection conn=JDBCConnectionFactory.getConnection();
    ChoiceQuestion choice=null;

    //查询该选择题
    try {
        Statement stmt=conn.createStatement();
        ResultSet rs=stmt.executeQuery("select *  from choiceQuestion "+"
            where cid='"+cid+"'");
        if(rs.next()){
            choice=new ChoiceQuestion (rs.getInt(1),rs.getString(2),rs.
                getString(3),rs.getString(4),rs.getString(5),rs.getString(6),
                rs.getString(7));
        }
    } catch(SQLException ee){
        ee.printStackTrace();
```

```java
            } finally {
                //关闭连接,略
            }
            return choice;
        }
        //插入一条填空题记录;返回值为 0 表示插入成功;返回值为-1 表示插入数据不成功
        public int insertChoiceQuestion(ChoiceQuestion choice){
            Connection conn=JDBCConnectionFactory.getConnection();
            if(selectOneChoiceQuestion(choice.getC_id())!=null)
                return -1;
            try {
                //插入一条记录
                String str=" insert into choicequestion (c_question,c_choiceA,c_
                    choiceB,c_choiceC,c_choiceD,c_answer)values('"
                    +choice.getC_question()+"','"+choice.getC_choiceA()+"','"
                    +choice.getC_choiceB()+"','"+choice.getC_choiceC()+"','"
                    +choice.getC_choiceD()+"','"+choice.getC_answer()+"')";
                PreparedStatement pstmt=conn.prepareStatement(str);
                if(pstmt.executeUpdate()==1){
                    return 0;
                } else { return -1;
                }
            } catch(SQLException ee){
                return -1;
            } finally {
                //关闭连接,略
            }
        }
    }
```

6) FillQuestionDAO 填空题数据访问类

```java
package dao.exam;

import java.sql.*;
import java.util.*;
import entity.exam.FillQuestion;

public class FillQuestionDAO {
    //产生多个不同的随机正整数的方法
    private int[] random_number(int max,int num){
        //同 ChoiceQuestionDAO 中的 random_number 方法,略
    }
    //从题库中随机生成填空题,返回保存 num 个填空题的链表
    public List<FillQuestion>selectFillQuestion(int num){
```

```
                //与 ChoiceQuestionDAO 中的 selectChoiceQuestion 方法类似,请参照完成
        }

        //根据题号查询一道填空题
        public FillQuestion selectOneFillQuestion(int fid){
                //与 ChoiceQuestionDAO 中的 selectOneChoiceQuestion 方法类似,请参照完成
        }

        //插入一条填空题记录;返回值为 0 表示插入成功;返回值为-1 表示插入数据不成功
        public int insertFillQuestion(FillQuestion fill){
                ////与 ChoiceQuestionDAO 中的 insertChoiceQuestion 方法类似,请参照完成
        }
}
```

以上数据访问类定义完成后,应测试每个类的数据访问方法,确保能够正常运行之后再进行下一步视图层 GUI 程序的编写。

4.3.4　GUI 界面类的实现

GUI 界面类主要实现用户与系统的交互界面,除定义基本的界面元素并进行布局之外,还要处理用户对界面的操作事件,即在事件处理方法中调用相应的数据访问类和方法,完成业务逻辑的处理。

在 Web 模式的项目中,通常视图层的界面可以由 Web 服务器端的网页实现,如 JSP 页面,而客户端只需要通过统一的浏览器访问服务器端页面。在 JSP 页面中嵌入调用业务逻辑方法或数据访问方法的代码;或者由 Servlet 来控制何时调用业务逻辑方法,以及如何显示 JSP 页面给客户端,从而实现 MVC 三层架构模式的程序。

1. 新建 GUI 图形用户界面类所在包

在项目文件夹下新建包 gui.exam,项目结构如图 4.31 所示。

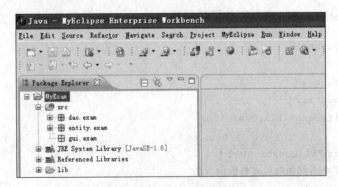

图 4.31　MyExam 项目结构

2. 新建 GUI 图形用户界面类

在 gui.exam 包下创建用户登录界面、"关于"界面、开始考试界面、考生考试界面、教

师管理界面和系统管理界面类。

3. 用户登录界面 LoginFrame

实现考生、教师、管理员用户登录功能,提供选择登录的用户类型功能;界面效果如图 4.32 所示。

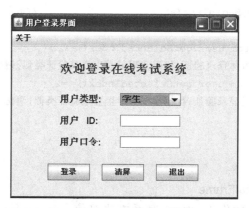

图 4.32　用户登录界面

实现该界面的按钮事件处理方法关键代码如下:

```
public void actionPerformed(ActionEvent e){
    if(e.getSource().equals(jb_clear)){           //清屏动作按钮处理
        jtf_id.setText("");
        jpf_password.setText("");
    }else if(e.getSource().equals(jb_exit)){      //退出动作按钮处理
        System.exit(0);
    }else{                                         //登录动作按钮处理
        String userId=jtf_id.getText().trim();
        String pwd=new String(jpf_password.getPassword()).trim();

        //验证用户名和密码不能为空
        if(userId.length()==0){
            JOptionPane.showMessageDialog(null,"请输入用户ID!","用户输入",1);
            jpf_password.setText("");
        }else if(pwd.length()==0){
            JOptionPane.showMessageDialog(null,"请输入用户口令!","用户输入",1);
            jtf_id.setText("");
        }else if(jcom_user.getSelectedIndex()==0){
            //学生用户登录验证
            StudentDAO sdao=new StudentDAO();
            Student st=sdao.selectStudent(userId);  //查询该学生信息
            if(st==null){
                JOptionPane.showMessageDialog(null,"该学生不存在!");
                jtf_id.setText("");jpf_password.setText("");
```

```
            }else if(!st.getPassword().equals(pwd)){
                JOptionPane.showMessageDialog(null,"口令错误!");
                jpf_password.setText("");
            }else{                                      //登录验证成功
                this.setVisible(false);
                new BeginTest(st);                      //显示开始考试界面
            }
        }else if(jcom_user.getSelectedIndex()==1){
            //教师登录验证,代码实现与学生登录验证类似,请仿照完成
        }else if(jcom_user.getSelectedIndex()==2){
            //管理员登录验证,代码实现与学生登录验证类似,请仿照完成
        }
    }
}
```

4."关于"界面 AboutFrame

实现显示系统开发的作者、地址、联系方式等信息功能,界面效果如图 4.33 所示。

此界面不涉及事件处理,实现比较简单,代码略。

5. 开始考试界面 BeginTest

实现考生登录后显示个人信息并开始计时考试功能,界面效果如图 4.34 所示。

单击"开始考试并开始计时"按钮,则显示消息对话框,提示本次考试的时间和题目数量分值等,如图 4.35 所示。

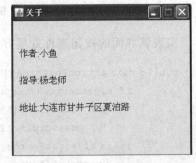

图 4.33 "关于"界面

图 4.34 开始考试界面

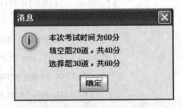

图 4.35 提示考试信息

开始考试界面的按钮动作事件处理方法关键代码如下:

```
public void actionPerformed(ActionEvent e){        //动作按钮监听方法实现
    if(e.getSource().equals(jb_exam)){             //"开始考试并开始计时"按钮
        JOptionPane.showMessageDialog(null,"本次考试时间为 60 分\n 填空题 20 道,"
```

```
            +"共 40 分\n 选择题 30 道,共 60 分");
        this.setVisible(false);
        new StudetFrame(st);                          //打开"学生管理"界面
    }else{                                            //"退出"按钮
        int temp = JOptionPane.showConfirmDialog(null,"您确定要取消此次考
        试吗?");
        if(temp==0)
            System.exit(0);                           //退出
    }
}
```

6．考生管理界面 StudetFrame

实现考生在线答题功能,选择题目界面效果如图 4.36 所示。

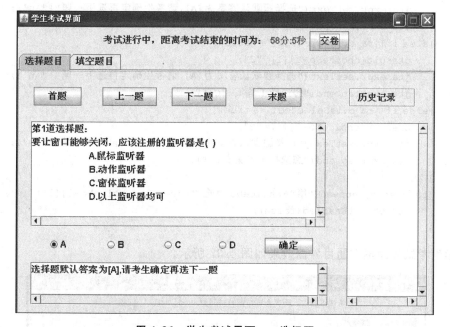

图 4.36　学生考试界面——选择题

选择题界面上按钮动作事件处理的关键代码如下：

```
//首题、上一题、下一题、末题按钮的动作事件处理
if(e.getSource().equals(jbc_first)){              //"首题"按钮
    jra_a.setSelected(true);                      //选项 A 默认选中
    jbc_ok.setEnabled(true);                      //设置确认按钮为可用
    jbc_priv.setEnabled(false);                   //设置"前一题"按钮不可用
    jbc_next.setEnabled(true);                    //"下一题"按钮可用
    jbc_last.setEnabled(true);                    //"末题"按钮可用
    showChoice(1);          //showChoice(int)方法在选择题界面上显示特定题号的试题
} else if(e.getSource().equals(jbc_priv)){        //上一题
    //与"首题"类似,参照上面补充代码
```

```java
    } else if(e.getSource().equals(jbc_next)){          //下一题
        //与"首题"类似,参照上面补充代码
    } else if(e.getSource().equals(jbc_last)){          //最后一题
        //与"首题"类似,参照上面补充代码
    }
//确定按钮的动作事件处理,将用户选择的答案显示在历史记录文本框中
if(e.getSource().equals(jbc_ok)){                       //"确定"按钮
    //保存当前用户所选答案
    if(jra_a.isSelected()==true){
        cresult[choiceNo-1][1]="A";
        jta_expr.setText("选择题默认答案为[A],请考生确定再选下一题"+'\n');
        jta_expr.append("您选择了答案:[A]");
    }else if(jra_b.isSelected()==true){
        cresult[choiceNo-1][1]="B";
        jta_expr.setText("选择题默认答案为[A],请考生确定再选下一题"+'\n');
        jta_expr.append("您选择了答案:[B]");
    }else if(jra_c.isSelected()==true){
        cresult[choiceNo-1][1]="C";
        jta_expr.setText("选择题默认答案为[A],请考生确定再选下一题"+'\n');
        jta_expr.append("您选择了答案:[C]");
    }else if(jra_d.isSelected()==true){
        cresult[choiceNo-1][1]="D";
        jta_expr.setText("选择题默认答案为[A],请考生确定再选下一题"+'\n');
        jta_expr.append("您选择了答案:[D]");
    }
    jta_history.append("第"+choiceNo+"道:"+cresult[choiceNo-1][1]+'\n');
    jbc_ok.setEnabled(false);
}
```

学生考试界面填空题目界面效果如图 4.37 所示。

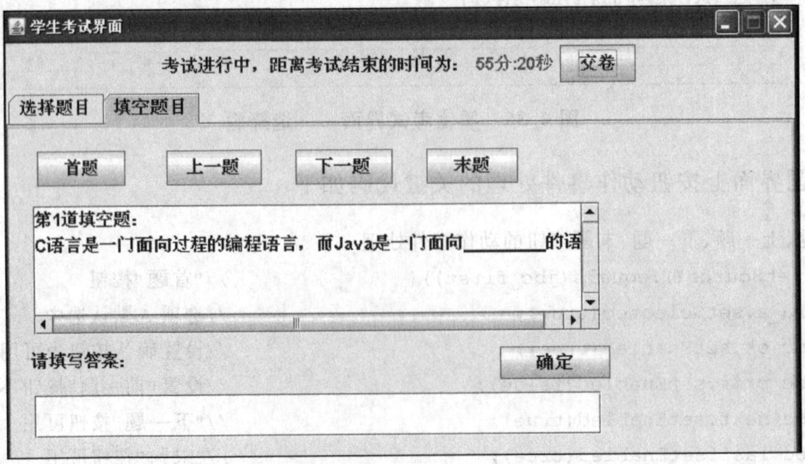

图 4.37 学生考试界面——填空题

填空题界面按钮动作事件处理的关键代码如下：

```
//以下处理填空题界面事件
    if(e.getSource().equals(jbf_first)){        //首题
        /* 单击首题、上一题、下一题、末题按钮的动作事件处理与选择题界面中相似,
         * 请仿照上面的程序完成,此处代码略
         */

            ⋮
        showFill(1);                            //在本界面显示填空题
    } else if(e.getSource().equals(jbf_priv)){  //上一题
        //略
    } else if(e.getSource().equals(jbf_next)){  //下一题
        //略
    } else if(e.getSource().equals(jbf_last)){  //末题
        //略
    }
    if(e.getSource().equals(jbf_ok)){           //确定按钮
        String temp=jta_result.getText().trim();
        fresult[fillNo-1][1]=temp;
        jbf_ok.setEnabled(false);
        jta_result.setText("");
    }
```

考试过程中,学生可以随时单击"交卷"按钮结束考试,界面如图 4.38 所示;单击"是"按钮,提示学生是否退出考试,界面如图 4.39 所示。

图 4.38　确认交卷

图 4.39　确认退出考试

"交卷"按钮事件处理的关键代码如下：

```
//"交卷"按钮事件处理
    if(e.getSource().equals(jb_exit)){
        int temp=JOptionPane.showConfirmDialog(null,"你确定要交卷吗？");
        if(temp==0){
            int score=0;
            for(int i=1; i<=30; i++){
                if(cresult[i-1][1].equalsIgnoreCase(cresult[i-1][0])&&cresult
                    [i-1][1].length()!=0){
                    score+=2;                   //每道选择题正确,总分加 2 分
                }
```

```
        }
        for(int i=1; i<=20; i++){
            if(fresult[i-1][1].equalsIgnoreCase(fresult[i-1][0])&&fresult
            [i-1][1].length()!=0){
                score+=2;                              //每道选择题正确,总分加2分
            }
        }
        int i=JOptionPane.showConfirmDialog(null,"你的得分是:"+score+"\n
        退出考试吗?");
        if(i==0){
            System.exit(0);
        }
    }
```

7. 教师管理界面 TeacherFrame

实现教师在线管理试题、修改个人密码功能,界面效果如图 4.40~图 4.42 所示。

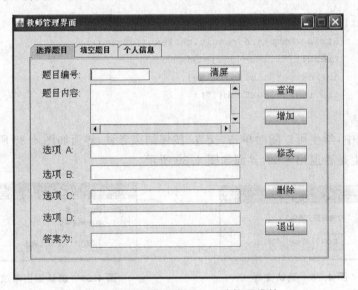

图 4.40 教师管理界面——选择题维护

教师管理选择题界面按钮"查询"、"增加"、"修改"、"删除"完成对选择题的管理,部分按钮的事件处理代码如下:

```
if(e.getSource().equals(jb_select)){            //选择题界面"查询"
    String cid=jtf_choice_id.getText().trim();  //获取题号
    if(cid.length()==0){                        //题号为空
        JOptionPane.showMessageDialog(null,"请输入题目编号!","提示信息",1);
    } else {                                    //执行查询并将结果显示在界面元素中
        try {
            ChoiceQuestion choice=cdao.selectOneChoiceQuestion
            (Integer.parseInt(cid));
```

```java
                    if(choice !=null){
                        jta_choice_text.setText(choice.getC_question());
                        jtf_a.setText(choice.getC_choiceA());
                         jtf_b.setText(choice.getC_choiceB());
                        jtf_c.setText(choice.getC_choiceC());
                        jtf_d.setText(choice.getC_choiceD());
                        jtf_result.setText(choice.getC_answer());
                    }
                } catch(NumberFormatException e2){
                    JOptionPane.showMessageDialog(null, "输入的题号不存在,请重新输入!");
                    jtf_choice_id.setText("");
                }
            }
    } else if(e.getSource().equals(jb_add)){           //选择题界面"增加"
        try {
            int cid=Integer.parseInt(jtf_choice_id.getText().trim());
            String cquestion=jta_choice_text.getText();
            String cchoiceA=jtf_a.getText();
            String cchoiceB=jtf_b.getText();
            String cchoiceC=jtf_c.getText();
            String cchoiceD=jtf_d.getText();
            String cresult=jtf_result.getText();
            int i=cdao.insertChoiceQuestion(new ChoiceQuestion(cid,
                    cquestion, cchoiceA, cchoiceB, cchoiceC, cchoiceD,cresult));
            if(i==0){                                  //函数返回值 0
                JOptionPane.showMessageDialog(null, "成功插入数据!");
                jtf_choice_id.setText("");
                jta_choice_text.setText("");
                jtf_a.setText("");           jtf_b.setText("");
                jtf_c.setText("");           jtf_d.setText("");
                jtf_result.setText("");
            } else {
                JOptionPane.showMessageDialog(null, "题号已存在,请选择其他编号!","提
                    示信息", 1);
                jtf_choice_id.setText("");
            }
        } catch(NumberFormatException e1){
            JOptionPane.showMessageDialog(null, "输入的题目编号错误,请重新输入!");
            jtf_choice_id.setText("");
        }
    } else if(e.getSource().equals(jb_alter)){         //选择题界面"修改"
        //略
    } else if(e.getSource().equals(jb_delete)){        //选择题界面"删除"
        //略
```

```
        } else if(e.getSource().equals(jb_exit)){         //选择题界面"退出"
            System.exit(0);
        }else if(e.getSource().equals(jb_clear)){         //选择题界面"清屏"动作按钮
            //略
        }
```

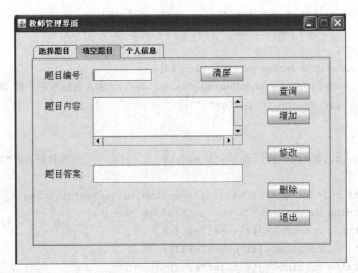

图 4.41　教师管理界面——填空题维护

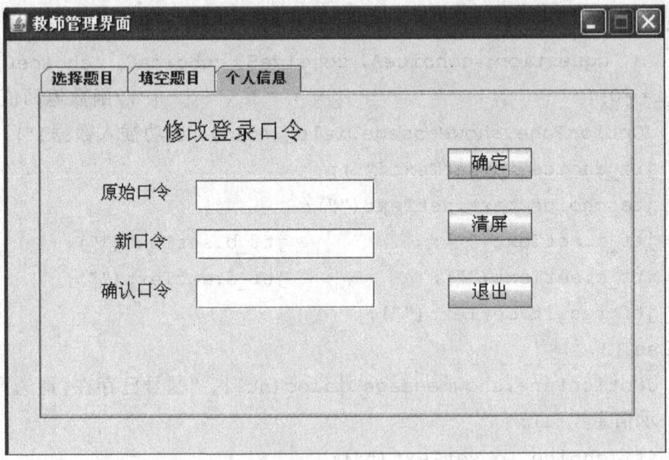

图 4.42　教师管理界面——个人信息

教师管理填空题界面的"查询"、"增加"、"修改"、"删除"完成对填空题的管理，其事件处理代码与选择题界面基本一致，请参考前面的程序完成本部分代码的编写：

```
        if(e.getSource().equals(jbf_clear)){              //填空题界面"清屏"
            jtf_fill_id.setText("");
            jtf_fill_result.setText("");
            jta_fill_text.setText("");
```

```java
    } else if(e.getSource().equals(jbf_select)){        //填空题界面"查询"
        //略
    } else if(e.getSource().equals(jbf_add)){           //填空题界面"增加"
        //略
    } else if(e.getSource().equals(jbf_alter)){         //填空题界面"修改"
        //略
    } else if(e.getSource().equals(jbf_delete)){        //填空题界面"删除"
        //略
    } else if(e.getSource().equals(jbf_exit)){          //填空题界面"退出"
        System.exit(0);
    }
```

教师管理个人信息界面完成教师用户个人密码的修改功能,主要的按钮事件处理代码如下:

```java
if(e.getSource().equals(jbi_ok)){
    String old=jpf_old.getText().trim();              //取出用户输入的口令
    String neew=jpf_new.getText().trim();
    String new_ok=jpf_new_ok.getText().trim();
    if(!old.equals(te.getPassword())){                //原口令输入不正确
        JOptionPane.showMessageDialog(null, "原始口令不正确!");
        jpf_old.setText("");
    } else if(!neew.equals(new_ok)){                  //新口令两次输入不同
        JOptionPane.showMessageDialog(null, "新口令不一致!");
        jpf_new.setText("");jpf_new_ok.setText("");
    } else {                                          //可以修改
        te.setPassword(new_ok);
        tdao.updateTeacher(te);
        JOptionPane.showMessageDialog(null, "成功修改口令!");
        jpf_old.setText("");jpf_new.setText("");jpf_new_ok.setText("");
    }
} else if(e.getSource().equals(jbi_clear)){           //清屏按钮
    //略
} else if(e.getSource().equals(jbi_exit)){            //退出按钮
    System.exit(0);
}
```

8. 系统管理界面 AdminFrame

实现管理员对学生、教师用户的管理和修改个人密码功能,系统管理界面由学生管理选项卡、教师管理选项卡和个人信息选项卡组成。

学生管理选项卡界面完成学生用户的查询、增加、删除、修改功能,界面效果如图 4.43 所示,教师管理选项卡界面完成教师用户的查询、增加、删除、修改功能,界面效果如图 4.44 所示。

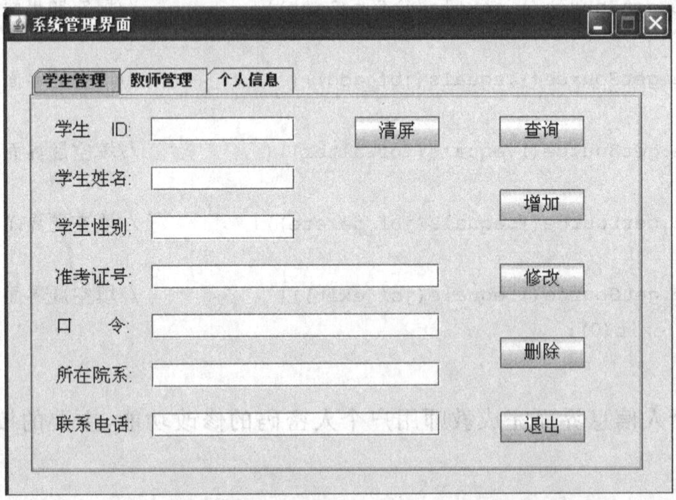

图 4.43　系统管理界面——学生信息管理

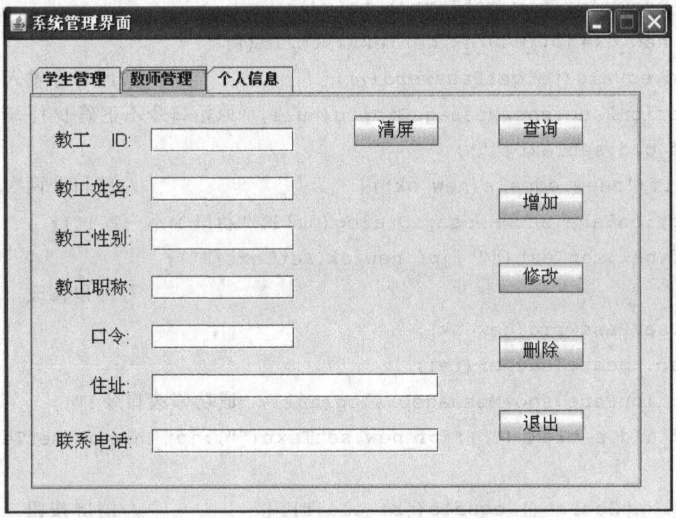

图 4.44　系统管理界面——教师信息管理

两个界面的按钮事件处理方法与前面的教师管理界面选择题选项卡按钮事件处理方法基本一致,请参考前面的程序完成程序代码片段：

```
//---------- 学生管理动作按钮----------------------------//
if(e.getSource().equals(jbs_clear)){                //学生管理清屏
    jtfs_id.setText("");    jtfs_name.setText("");    jtfs_sex.setText("");
jtfs_zkzh.setText("");    jtfs_code.setText("");    jtfs_dep.setText("");
    jtfs_phone.setText("");
} else if(e.getSource().equals(jbs_select)){        //学生管理查询
    //略
} else if(e.getSource().equals(jbs_add)){           //学生管理增加
    //略
```

```
    } else if(e.getSource().equals(jbs_alter)){        //学生管理修改
        //略
    } else if(e.getSource().equals(jbs_delete)){       //学生管理删除
        //略
    } else if(e.getSource().equals(jbs_exit)){         //学生管理退出
        System.exit(0);
    }
//-----------教师管理动作按钮--------------------------------//
    if(e.getSource().equals(jbt_clear)){               //教师管理界面"清屏"按钮
        //略
    } else if(e.getSource().equals(jbt_select)){       //教师管理界面"查询"按钮
        //略
    } else if(e.getSource().equals(jbt_add)){          //教师管理界面"增加"按钮
        //略
    }else if(e.getSource().equals(jbt_alter)){         //教师管理界面"修改"按钮
        //略
    } else if(e.getSource().equals(jbt_delete)){       //教师管理界面"删除"按钮
        //略
    } else if(e.getSource().equals(jbt_exit)){
        System.exit(0);
    }
```

个人信息选项卡界面实现管理员用户口令的修改,界面效果如图 4.45 所示。

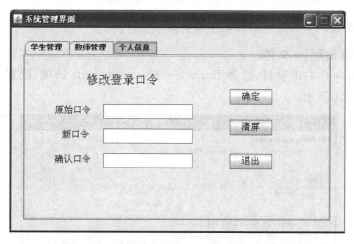

图 4.45　系统管理界面——修改管理员口令

其按钮动作事件处理与教师管理界面的"个人信息"选项卡功能类似,事件处理方法也基本一致,请参考前面的程序完成代码:

```
//--------个人信息 动作按钮-------------------------
    if(e.getSource().equals(jbi_ok)){                  //"确定"按钮
        //略
    } else if(e.getSource().equals(jbi_clear)){        //"清屏"按钮
        //略
```

```
        } else if(e.getSource().equals(jbi_exit)){            //"退出"按钮
            System.exit(0);
        }
```

至此，Java 在线考试系统实现全部完成，在 MyEclipse 环境中单个类文件测试全部通过的情况下，完成系统的整合和运行。请设计测试用例并完成系统测试，以检验和修正考试系统项目中的逻辑错误。

4.4 项目发布与改进

4.4.1 项目发布

在 MyEclipse 开发环境完成考试系统项目的开发和测试运行后，可以进行项目的发布，使项目在 Windows 操作系统环境下正常运行。

1. 确保数据库访问正常

MySQL 服务器正常运行，考试系统数据库 exam 已正确建立或附加到数据库服务器上，假设 MySQL 数据库服务器地址为 192.168.1.100，请修改 DAO 数据访问包中的 JDBCConnectionFactory 数据库连接工厂类里定义的数据库连接字符串为：

```
String url="jdbc:mysql://192.168.1.100:3306/Exam";
```

确保能够连接到系统统一的数据库服务器。

2. 将项目导出到 jar 文件

在 MyEclipse 中右击项目，选择 Export…→Java→Jar file 选项，设置导出文件的存放位置，如图 4.46 所示。

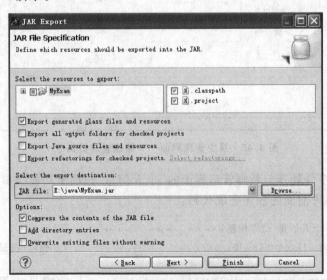

图 4.46 导出项目

先后两次单击 Next 按钮,选择项目主文件,如图 4.47 所示。

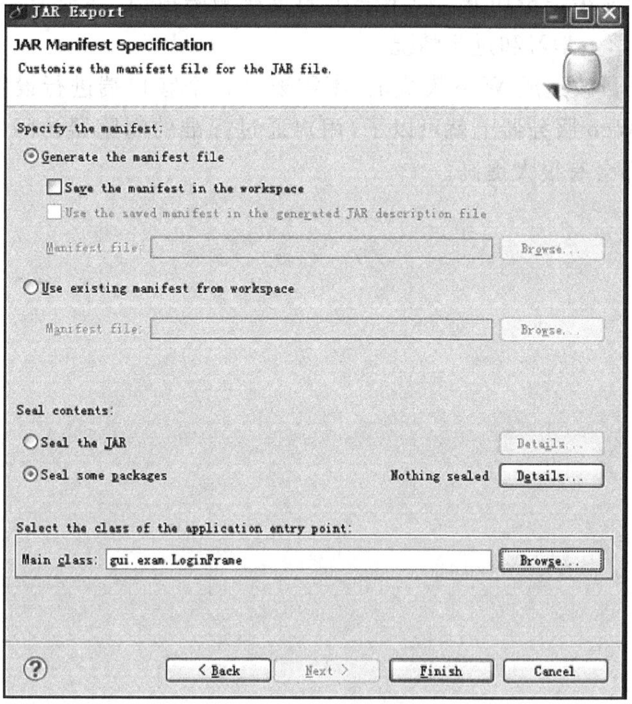

图 4.47 选择项目主文件

单击 Finish 按钮完成导出,生成可以自运行的 Exam.jar 文件就可以运行了。

3. 生成批处理文件或可执行文件

编辑项目压缩文件 META-INF 目录下 MANIFEST.MF 文件,确保文件中有以下三条语句:

```
Manifest-Version: 1.0
Created_By: 1.7.0
Main-Class: gui.exam.LoginFrame
```

重新打包后,会生成压缩文件 Exam.rar。

新建一个批处理文件 exam.bat(文件名任意),编辑文件内容包含语句:

```
javaw-jar MyExam.rar
```

运行批处理文件,就可以运行考试系统项目了。

4.4.2 项目改进

上述的 Java 在线考试系统程序基本实现了项目需求,在此基础上还可以考虑从以下几个方面进行改进。

(1) 为程序增加试题组卷存储功能:可以由教师在线选择试题类型和数量,随机组卷,并保存在试卷数据库中,供同一层次学生考试使用。

(2) 增加考生成绩记录功能:记录学生的每次测验成绩,并形成一个对比数据和趋势表,有利于掌握学生学习和进步情况。

(3) 可以将项目改造成 Web 模式的,不需要在每个客户端进行设置和操作,只需把项目代码发布在 Web 服务器上就可以了,用户通过标准的浏览器访问考试系统,系统维护量和用户体验都会有很大提高。

第 5 章 网上灯饰店的研究与实现

本章介绍的网上灯饰店不是一个实际的建设项目,系统目标和功能是研究人员参照一般网店、结合教学过程中的实际情况而设定的。该网店的功能虽然不是很全面,可能还存在某些缺陷,但网店建设是按照软件工程的思想、基本按照生命周期法的开发过程进行的,过程描述清晰,建设网店过程中遇到的核心问题、主要问题,都能在本章找到解决方法。

本章介绍的网上灯饰店建设是基于 Internet 采用 MVC 框架的,其软件开发环境如下:

(1) 操作系统:Windows XP。
(2) 开发环境:Visual Studio 2010(采用 ASP.NET 的 C♯编程)。
(3) 数据库:SQL Server 2008。

5.1 网上灯饰店规划

在系统规划阶段主要完成了如下工作:
(1) 简述网店的发展和建设网店的意义;
(2) 网店需求分析和目标设定;
(3) 对设定的目标和需求进行可行性分析。

5.1.1 网上商店系统发展和实现网上商店系统的意义

近几年来,电子商务在国内的高速发展带动了为电子商务服务的软件行业的发展,随之而来的是诞生了许多与电子商务密切相关的网上商店系统和网上商城系统。对于网上商店的实现技术,无论是开源的还是商业的,都非常多。网上商城系统,指的是支持多商户的大卖场模式,如淘宝商城、京东商城等,它们在利用互联网做网上商城方面就取得了很大的成功;而网上商店,则一般指的是独立 B2C 网站。在软件架构上,两者也是有所不同的,即一个是支持多租户模式的,一个是不支持多租户模式的。

与传统商店比较,网上商店是一个虚拟的商店,具有许多新的特点以及其特有的优势。如网上商店的运营成本比传统商店的运营成本要低很多,首先,它不需要支付高额的店铺租金以及装修费。其次,它也不需要将大笔的流动资金作为货物而沉淀在店里,它可以依靠"零存储"管理,缩短资金的周转周期。同时网上商店更容易开拓市场,利用网络全球化的优势打开国内外市场。这些特性使网上商店拥有很强的生命力,使其成为传统商业的有力挑战者。网上商店将会是传统商店的一种更新、改造、提升的网络平台。网上商店的发展趋势应该是网上商店和传统商店合作,这样就能够发挥各自的优势,实现优势互补,能使商家在这种互补的新型经营模式下取得更大的利益。

在市场竞争如此激烈的环境下,各种生产商、制造商、零售商都需要开拓新的销售渠道来推销自己的产品,扩展自己品牌的市场。网上商店系统可以帮助商家以更节省财力、更高效的方式推销出自己的商品,获得高效带来的更多利益。所以他们都纷纷抓住机遇,以网络作为销售渠道,先后都拥有了自己的网上商店系统。而对于消费者来说,进行网上购物是一件轻松、快捷、方便的购物方式,与传统购物模式相比之下,网购可以省去很多麻烦,可以不受时间、地点的限制,只需连网登录网上商店系统就可以轻松购买心仪物品,而且可以很轻松地做到货比三家,既省时省力又省钱的网购方式就成了众多消费者的选择。所以说开发网上商店系统是很有实际意义的。

5.1.2 网上灯饰店的需求分析

建立网上灯饰店首先要根据用户对需求的描述,结合实际情况和未来进一步的发展,定义出网上灯饰店的建设目标和需求。

网上灯饰店的建设目标如下:
(1) 制定出规范、合理的业务流程;
(2) 为高层管理者提供决策支持;
(3) 提高销售效率和销售量;
(4) 加强对整个销售过程的监控力度,保证系统和数据安全;
(5) 实现各类信息的集中化、数字化、规范化管理。

网上灯饰店的需求如下:
(1) 建立一个相对完善的网店;
(2) 记载的信息准确、规范,方便使用;保证信息的安全;
(3) 功能全面;
(4) 方便各类人员完成相关操作,各种操作合乎用户的操作习惯;
(5) 为相关的人员提供进行决策的有效数据;
(6) 提高销售量。

为了实现目标和满足需求网上灯饰店的应具备的功能如下。
(1) 用户管理。建立完善的不同用户身份的权限管理:不同权限拥有不同功能和操作权限,以便更好地管理用户、保障网站能更好地运行;用户可以在线注册,成为网站会员,享有相关的权益、履行相关义务。
(2) 商品管理。网店操作员可以通过浏览器在线添加、修改、删除商品信息,商品的数据可以即时更新,保证用户浏览到最新的商品信息;用户能够对商品进行查询、购买或收藏自己喜欢的商品。

(3)订单管理。实现不同权限的人员对订单完成不同的管理,如操作员可以通过浏览器查看各用户(普通用户)的订单状态、执行订单等,普通用户可以查看自己的订单、删除未结算的订单、进行结算等。

5.1.3 网上灯饰店可行性研究

对定义的目标和需求进行经济、技术、社会三方面可行性研究。

1. 经济可行性分析

对系统的经济可行性进行具体分析如下。

1)系统初期投资

建立网店的硬件和软件支持费用估计需要 2.5 万元,对网店建设的人工初期投资计算如表 5.1 所示。

表 5.1 网上灯饰店系统初期投资

序号	项目	人工/(人·日)	单价/元	合计/千元
1	规划和分析	15	200	3
2	总体设计	15	200	3
3	用户管理子系统详细设计	5	200	1
4	商品管理子系统详细设计	5	200	1
5	订单管理子系统详细设计	10	200	2
6	结算子系统详细设计	10	200	2
7	购物车子系统详细设计	5	200	1
8	我的收藏子系统详细设计	5	200	1
9	数据库的设计与实现	10	200	2
10	编码实现	30	200	6
11	单体测试	10	200	2
12	系统集成测试	10	200	2
13	说明手册编制	5	200	1
14	合计			27

最初投资额=硬件和软件购置费+人工初期投资=5.2 万元。

2)系统货币的时间价值

对网店建设的货币的时间价值计算如表 5.2 所示。

表 5.2　网上灯饰店系统的货币时间价值

年份	将来值/万元	$(1+i)^n$	现在值/万元	累计的现在值/万元
1	3	1.05	2.857	2.857
2	3	1.1025	2.721	5.578
3	3	1.1576	2.591	8.169
4	3	1.2155	2.468	10.637
5	3	1.2763	2.351	12.988

3) 投资回收期

引入网上灯饰店系统一年后,可节省 2.857 万元,比最初投资额少 2.343 万元。但第二年可节省 2.721 万元。即

$$投资回收期:1+2.343/2.721=1.861(年)$$

运行本网店的硬件和软件支持费用估计需要 2.5 万元,软件开发费用估计 2.7 万元,在系统投入使用后,每年可为商店节省支出约 3 万元,假设本软件可以使用 5 年。一次性支出为 5.2 万元,现假定这笔投资费用存入银行的利息按 3.5% 计算,共计 $=5.2\times(1+3.5\%)^5=6.176$ 万元,纯收入为:12.988－6.176＝6.812(万元)(注:运行期间的人工费同不采用系统情况的人工费相抵)。

经过以上的分析,此网上灯饰店系统在经济上的开发是可行的。

2. 技术可行性分析

ASP.NET 技术、开发平台 Visual Studio 2010 以及后台数据库 SQL Server 2008 都是很成熟的技术和开发环境,完全满足开发灯饰店的需求。

这些技术和开发环境对开发人员技术的要求不是很高;并且网上有非常丰富的相关资料可供开发人员参考和利用。因此,开发灯饰店在技术上是可行的。

3. 社会可行性分析

本灯饰店属于教学案例本学校自行开发的应用系统,与现存的网上商店系统有所不同,所以不存在侵权问题。而且也没有涉及到侵犯个人隐私或个人责任方面的问题,故使用本系统不会引发法律方面的纠纷问题,从而开发本网店具有社会可行性。

综上所述,本系统的开发是可行的。

5.2　网上灯饰店分析

在系统分析阶段主要完成如下工作。

1. 业务流程分析与描述

分析和描述的主要业务如下:

(1) 用户管理;

(2) 订单管理;

(3) 商品管理。

描述工具是业务流程图。

2. 数据流程分析与描述

利用数据流程图描述了如下数据流程：

(1) 网店顶层数据流程；

(2) 网店第一层数据流程；

(3) 网店第二层中的用户信息管理数据流程；

(4) 网店第二层中的订单信息管理数据流程；

(5) 网店第二层中的商品信息管理数据流程。

3. 数据分析与描述

利用表格对各数据项、外部实体、各数据流、各数据存储以及各加工处理进行了描述，即建立了数据流图中各种对象的数据字典。

5.2.1 业务流程分析与描述

根据网店的目标和需求，将网店的基本业务分成三大业务：第一是用户信息管理业务，主要是对用户进行管理，如用户注册、用户登录、用户信息查询、用户信息修改、删除用户等；第二是订单管理业务，主要是对订单的基本信息进行管理，如生成订单，订单信息查询，删除、修改订单信息，订单结算等；第三是商品管理业务，主要是对商品的一些基本信息进行管理，如商品信息查询、收藏商品、将商品加入购物车和添加商品等。

1. 用户信息管理业务流程分析与描述

用户首先要进行注册才能进行相关操作，已注册的用户登录后进行相关操作，主要操作有基本信息进行查询、修改。操作员可以进行信息查询、修改和删除长时间不登录系统的用户。用户信息管理业务流程图如图 5.1 所示。

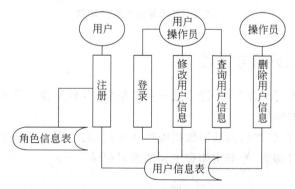

图 5.1 用户管理业务流程图

2. 商品信息管理业务分析与描述

一般用户登录本网店后可以查看商品详细信息，根据需要按条件查询所需商品。注册登录系统后，可以将心仪的商品加入收藏夹或购物车，对收藏夹或购物车中的商品信息进行查询、修改或删除,购买商品(即生成订单)和对商品做出自己的评价。操作员登录后可以添加商品和修改商品,商品信息管理业务流程图如图5.2所示。

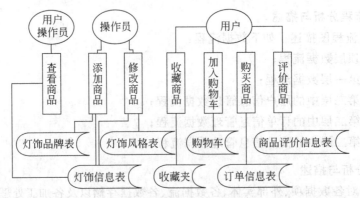

图 5.2 商品管理业务流程图

3. 订单信息管理业务

注册用户在登录系统后,可以查看自己所有的订单,修改订单信息、删除订单信息,然后对确定的订单进行结算操作,对收到货的订单进行确认收货操作完成一笔交易;操作员可以查看订单,对结账的订单进行发货。订单信息管理业务流程图如图5.3所示。

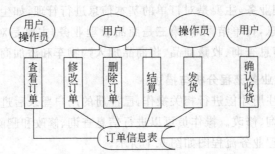

图 5.3 订单管理业务流程图

5.2.2 数据流程分析与描述

把各业务流程进行综合分析,进行进一步的抽象,分析数据的流向和处理,应用数据流程图描述出分析的结果。

首先分析、抽象出灯饰店的顶层数据流程和处理,通过图5.4表达出来,顶层数据流图中表达了抽象的灯饰店、外部实体和主要的数据存储。

图 5.4 灯饰系统店顶层数据流图

依据业务流程分析的结果、结合顶层数据流分析,把顶层数据流图中的处理——灯饰店系统进行功能分解,在功能分解的同时把数据流、数据存储和外部实体也都进行分解,对分解的结果进行描述,得到灯饰店的第一层数据流图,如图5.5所示。

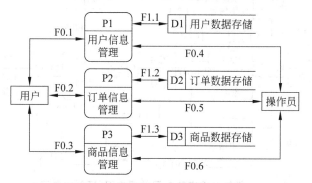

图5.5　灯饰系统店第一层数据流图

结合业务流程分析的结果、对第一层数据流图中的各处理进行进一步分解,同时分解相应的数据流、数据存储;第一层数据流图中的用户信息管理进行分解的结果通过图5.6描述出来;图5.7描述出第一层数据流图中的订单信息管理进行分解的结果;图5.8描述出第一层数据流图中的商品信息管理进行分解的结果。

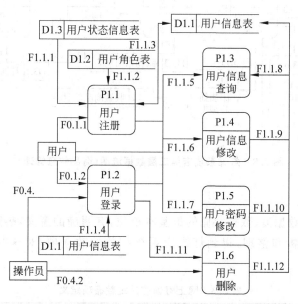

图5.6　灯饰系统店第二层数据流图(用户信息管理)

5.2.3　数据分析

对数据流图中的数据流、数据存储、处理和外部实体进行说明,并就数据流的组成和数据存储的组成进行说明,由数据项构成数据结构,再由数据结构和数据项得到数据流和存储数据。

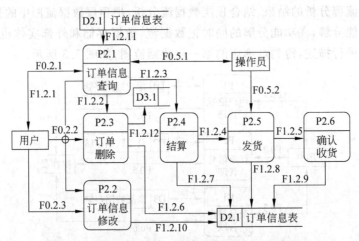

图 5.7 灯饰系统店第二层数据流图(订单信息管理)

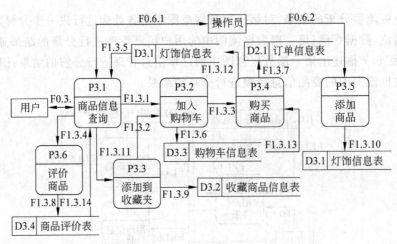

图 5.8 灯饰系统店第二层数据流图(商品信息管理)

1. 数据项定义

数据项是组成数据流、数据存储的最基本单位,有明确的意义,表达事物的属性和特征,经过调查、分析和整理,得到灯饰店系统相关的数据项,对这些数据项的描述如表 5.3 所示。

表 5.3 网上灯饰店系统数据项定义

编号	名 称	别 名	类 型	长度/字节
I01	UserId	用户编号	int	4
I02	UserName	用户姓名	Varchar	20
I03	Address	用户地址	Varchar	50
I04	PostalCode	邮编	Varchar	6

续表

编号	名称	别名	类型	长度/字节
I05	Phone	电话号码	Varchar	11
I06	Email	邮箱	Varchar	50
I07	Gender	性别	Varchar	4
I08	UserRoleId	用户角色编号	int	4
I09	UserStateId	用户状态编号	int	4
I10	LoginId	登录用户名	Varchar	50
I11	LoginPwd	登录密码	Varchar	50
I12	LastLoginTime	最后登录时间	Varchar	10
I13	LightId	灯饰编号	int	4
I14	CatagoryId	种类编号	int	4
I15	BrandId	品牌编号	int	4
I16	StyleId	风格编号	int	4
I17	Stuff	材质	Varchar	20
I18	Color	颜色	Varchar	20
I19	Locate	放置位置	Varchar	20
I20	Price	单价	decimal	18
I21	Quantity	库存量	int	4
I22	Description	商品描述	Varchar	250
I23	Images	图片地址	Varchar	20
I24	BrandName	品牌名称	Varchar	20
I25	CatagoryName	种类名称	Varchar	20
I26	StyleName	风格名称	Varchar	10
I27	CommentId	评价编号	int	4
I28	Comments	评价详情	Varchar	150
I29	Date	评价日期	Varchar	10
I30	ShopCartId	购物车商品编号	int	4
I31	AddDate	添加日期	Varchar	10
I32	OrderId	订单编号	Varchar	17
I33	Quantity	购买数量	int	4
I34	TotalPrice	总价	decimal	18
I35	Date	下订单日期	Varchar	10

续表

编号	名称	别名	类型	长度/字节
I36	OrderStateId	订单状态编号	int	4
I37	ShowOrNot	是否显示	int	1
I38	ReceiveName	收件人姓名	Varchar	20
I39	ReceivePhone	收件人电话	Varchar	11
I40	PostType	快递类型	Varchar	20
I41	PaymentType	支付类型	Varchar	20
I42	PostFee	快递价格	int	4
I43	OrderStateName	订单状态名	Varchar	20
I44	Clicks	点击次数	int	4
I45	StoreCount	收藏次数	int	4
I46	SaleCount	购买次数	int	4
I47	UserRoleName	用户角色名	Varchar	20
I48	WishListId	收藏编号	int	4
I49	StoreDate	收藏日期	Varchar	10
I50	UserStatesName	用户状态名	Varchar	20

2．数据结构

因为在不同的数据流和数据存储中包含相同的数据项，为了简化数据流和数据存储的描述，把若干相关数据项(即经常同时出现在组成数据结构、数据流和数据存储中的数据项)组成数据结构。经过分析和整理，设计出的主要数据结构如表5.4所示。

表5.4 网上灯饰店系统数据结构定义

序号	编号	名称	组成结构
1	S01	用户注册信息	I02,I11,I10,I07,I05,I06,I04,I03
2	S02	用户登录信息	I10,I11
3	S03	用户修改信息	I10,I05,I02,I03,I04,I06
4	S04	用户修改密码信息	I11
5	S05	用户-订单查询信息	I36
6	S06	订单删除信息	I32
7	S07	用户-结算信息	I32,I40,,I02,I05,I06,I03,I41
8	S08	用户-订单修改信息	I32,I33
9	S09	用户-确认收货信息	I32,I36

续表

序号	编号	名 称	组 成 结 构
10	S10	商品查询信息	查询条件
11	S11	购物车信息	I30,I13,I01,I31
12	S12	收藏夹信息	I48,I01,I49,I13
13	S13	操作员-发货信息	I32,I36
14	S14	操作员-商品查询信息	I32,I36
15	S15	商品(灯饰)信息	I13,I25,I24,I26,I17,I18,I19,I20,I21,I22
16	S16	用户状态信息	I50
17	S17	用户角色信息	I10
18	S18	用户信息	I01,I02,I11,I10,I07,I05,I06,I04,I03,I12
19	S19	订单信息	I01,I13,I33,I35,I38,I39,I03,I04,I05,I40,I41
20	S20	商品评价信息	I13,I01,I28,I29
21	S21	用户删除信息	I01
22	S22	订单修改信息	I32,I33,I13,I01
23	S23	购买数量	I13+I33

3．数据流

对数据流图中的各数据流进行整理和描述,主要描述内容包括数据流的编号、名称、来源、去向和组成,详细描述如表5.5所示。

表5.5 网上灯饰店系统数据流定义

序号	编号	名 称	来 源	去 向	组 成 结 构
1	F0	用户-系统信息	用户(P0)	P0(用户)	F0.1+F0.2+F0.3
2	F0.1	用户-用户管理信息	用户(P1)	P1(用户)	S01+S02
3	F0.1.1	用户注册信息	用户	P1.1	S01
4	F0.1.2	用户登录信息	用户	P1.2	S02
5	F0.2	用户-订单管理信息	用户(P2)	P2(用户)	S03+S05+S06
6	F0.2.1	用户-订单查询信息	用户(P2.1)	P2.1(用户)	S05
7	F0.2.2	用户-订单删除信息	用户	P2.2	S06
8	F0.2.3	用户-订单修改信息	用户	P2.3	S03
9	F0.3	用户-商品管理信息	用户(P3)	P3(用户)	S10
10	F0.3.1	用户-商品查询信息	用户(P3.1)	P3.1(用户)	S10
11	F0.4	操作员-用户管理信息	操作员(P1)	P1(操作员)	S18+S21

续表

序号	编号	名称	来源	去向	组成结构
12	F0.4.1	操作员-用户查询信息	操作员	P1.2	S18
13	F0.4.2	操作员-用户删除信息	操作员	P1.6	S21
14	F0.5	操作员-订单管理信息	操作员(P2)	P2(操作员)	S19+S13
15	F0.5.1	操作员-订单查询信息	操作员(P2.1)	P2.1(操作员)	S19
16	F0.5.2	操作员-发货信息	操作员	P2.5	S13
17	F0.6	操作员-商品管理信息	操作员(P3)	P3(操作员)	S14+S15
18	F0.6.1	操作员-商品查询信息	操作员(P3.1)	P3.1(操作员)	S14
19	F0.6.2	操作员-添加商品信息	操作员	P3.5	S15
20	F1	系统-存储信息	D0(P0)	P0(D0)	F1.1+F1.2+F1.3
21	F1.1	用户信息管理-用户信息存储	P1(D1)	D1(P1)	S02+S03+S04+S16+S17+S18+S21+S22
22	F1.1.1	用户状态信息	D1.3	P1.1	S16
23	F1.1.2	用户角色信息	D1.2	P1.1	S17
24	F1.1.3	用户注册信息	P1.1(D1.1)	D1.1(P1.1)	S18
25	F1.1.4	用户登录信息	D1.4	P1.2	S02,S22
26	F1.1.5	用户查询信息1	用户/P1.1	P1.3	S18
27	F1.1.6	用户修改信息1	用户/P1.1	P1.4	S03
28	F1.1.7	用户修改密码信息1	用户/P1.5	D1.1	S04
29	F1.1.8	用户查询信息2	D1.1	P1.3	S18
30	F1.1.9	用户修改信息2	P1.4	D1.1	S03
31	F1.1.10	用户修改密码信息2	P1.5(D1.1)	D1.1(P1.5)	S04
32	F1.1.11	用户删除信息1	P1.2	P1.6	S21
33	F1.1.12	用户删除信息2	P1.6	D1.1	S21
34	F1.2	订单信息管理-订单信息存储	P2(D2)	D2(P2)	S06+S07+S19+S22
35	F1.2.1	订单修改信息	P2.1	P2.2	S19
36	F1.2.2	订单删除信息	P2.1	P2.3	S06
37	F1.2.3	订单结算信息	P2.1	P2.4	S07
38	F1.2.4	已结算订单信息	P2.4	P2.5	S07
39	F1.2.5	已发货订单信息	P2.5	P2.6	S07
40	F1.2.6	订单删除信息	P2.3	D2.1	S06

续表

序号	编号	名 称	来 源	去 向	组 成 结 构
41	F1.2.7	订单结算信息	P2.4	D2.1	S07
42	F1.2.8	订单发货信息	P2.5	D2.1	S07
43	F1.2.9	订单确认收货信息	P2.6	D2.1	S07
44	F1.2.10	订单修改信息	P2.2	D2.1	S22
45	F1.2.11	订单查询信息	P2.2	D2.1	S19
46	F1.2.12	订单删除信息	P2.3	D3.1	S23
47	F1.3	商品信息管理-商品信息存储	P3(D3)	D3(P3)	S11＋S12＋S15＋S21
48	F1.3.1	用户欲购的灯饰信息	P3.1	P3.2	S12
49	F1.3.2	用户购买的灯饰信息	P3.3	P3.2	S11
50	F1.3.3	用户购买的灯饰信息	P3.2	P3.4	S15
51	F1.3.4	商品评价信息	P3.1	P3.6	S21
52	F1.3.5	灯饰信息	D3.1	P3.1	S15
53	F1.3.6	欲购商品购物信息	P3.2	D3.3	S11
54	F1.3.7	订购的商品信息	P3.4	D2.1	S15
55	F1.3.8	商品评价信息	P3.6	D3.4	S21
56	F1.3.9	欲收藏的商品信息	P3.3	D3.2	S12
57	F1.3.10	灯饰信息	P3.5	D3.1	S15
58	F1.3.11	用户收藏的灯饰信息	用户	P3.3	S12
59	F1.3.12	购买数量	P3.4	D3.1	S23
60	F1.3.13	欲购商品购物信息	D3.3	P3.4	S15
61	F1.3.14	商品评价信息	D3.4	P3.1	S21

4．数据存储

对数据流图中的各数据存储进行整理和描述，主要描述内容包括数据存储的编号、名称、输入数据流、输出数据流、作用简述等，详细描述如表5.6所示。

表5.6 网上灯饰店系统数据存储定义

序号	编 号	名 称	输入数据流	输出数据流	作用简述
1	D0	相关数据存储	F1	F1	存储灯饰店系统的相关数据
2	D1	用户数据存储	F1.1	F1.1	存储灯饰店系统的用户数据

续表

序号	编号	名称	输入数据流	输出数据流	作用简述
3	D1.1	用户信息表	F1.1.3, F1.1.9, F1.1.10, F1.1.11	F1.1.3, F1.1.4, F1.1.8	存储用户的基本信息
4	D1.2	用户角色表		F1.1.1	存储系统中用户分类信息
5	D1.3	用户状态信息表		F1.1.2	存储系统中用户状态信息
6	D2	订单数据存储	F1.2	F1.2	存储灯饰店系统的订单数据
7	D2.1	订单信息表	F1.3.7, F1.2.6, F1.2.7, F1.2.8, F1.2.9, F1.2.10	F1.2.11	存储订单的基本信息
8	D3	灯饰数据存储	F1.3	F1.3	存储灯饰店系统的灯饰数据
9	D3.1	灯饰信息表	F1.2.11, F1.3.10, F1.3.12	F1.3.5	存储灯饰的基本信息
10	D3.2	收藏商品信息表	F1.3.9	F1.3.2	存储用户收藏的灯饰信息
11	D3.3	购物车信息表	F1.3.6	F1.3.13	存储用户欲购的灯饰信息
12	D3.4	商品评价表	F1.3.8	F1.3.14	存储用户对灯饰的评价信息

5. 数据处理

对数据流图中的各数据处理进行整理和描述,主要描述内容包括数据处理的编号、名称、输入数据流、输出数据流、功能简述等,详细描述如表 5.7 所示。

表 5.7 网上灯饰店系统数据处理定义

序号	编号	名称	输入数据流	输出数据流	功能说明
1	P0	灯饰店系统	F0, F1	F0, F1	网上灯饰店
2	P1	用户信息管理	F0.1, F1.1	F0.1, F1.1	对用户信息进行管理
3	P1.1	用户注册	F0.1.1, F1.1.1, F1.1.2, F1.1.3	F1.1.3, F1.1.5, F.1.1.6, F1.1.7	完成用户注册
4	P1.2	用户登录	F0.1.2, F0.4.1, F1.1.4	F1.1.11, F1.1.5, F.1.1.6, F1.1.7	完成用户登录
5	P1.3	用户信息查询	F1.1.5, F1.1.8		查询用户基本信息
6	P1.4	用户信息修改	F1.1.6	F1.1.9	用户修改自己的基本信息
7	P1.5	用户密码查询	F.1.1.7	F.1.1.10	修改自己的登录密码
8	P1.6	用户删除	F1.1.11, F.0.4.2	F1.1.12	操作员删除三个月及以上没用登录的用户
9	P2	订单管理	F0.2, F1.2	F0.2, F1.2	对订单信息进行管理

续表

序号	编号	名称	输入数据流	输出数据流	功能说明
10	P2.1	订单信息查询	F0.2.1,F1.2.11,F0.5.1	F1.2.1,F1.2.2,F1.2.3	用户可以查自己的订单,管理员可以查用户的订单
11	P2.2	订单信息修改	F1.2.1,F.0.2.3	F1.2.10	用户对自己的订单进行修改
12	P2.3	订单删除	F0.2.2,F1.2.2,F1.2.6	F1.2.6,F1.2.12	用户删除自己的订单
13	P2.4	结算	F1.2.3	F1.2.4,F1.2.7	用户结算自己的订单(订单状态改变)
14	P2.5	发货	F0.5.2,F1.2.4	F1.2.5,F1.2.8	操作员对用户结算的订单进行发货(订单状态改变)
15	P2.6	确认收货	F1.2.5	F1.2.9	用户对收到货品的订单进行收货确认(订单状态改变)
16	P3	商品信息管理	F0.3,F1.3	F0.3,F1.3	对商品信息进行管理
17	P3.1	商品查询	F0.3.1,F0.6.1,F1.3.5F1.3.14	F1.3.2,F1.3.11	用户和操作员查询商品信息
18	P3.2	加入购物车	F1.3.1,F.1.3.2	F1.3.6,F1.3.3	用户把自己想购买的灯饰放入购物车中
19	P3.3	添加到收藏夹	F1.3.11	F1.3.2,F1.3.9	用户把自己感兴趣的灯饰进行收藏
20	P3.4	购买商品	F1.3.3,F1.3.13	F1.3.4,F1.3.7,F1.3.12	用户买下确定购买的灯饰,生成订单
21	P3.5	添加商品	F0.6.2	F1.3.10	操作员添加新灯饰,或修改原有灯饰
22	P3.6	评价商品	F1.3.4	F1.3.8	用户对商品进行评价

6. 外部实体

对数据流图中的各外部实体进行整理和描述,主要描述内容包括数据流的编号、名称和说明,详细描述如表5.8所示。

表5.8 网上灯饰店系统外部实体定义

序号	名称	说明	序号	名称	说明
1	用户	系统注册的用户	2	操作员	系统操作员

5.3 网上灯饰店设计

在系统设计阶段主要完成了如下工作:
(1)网上灯饰店功能结构设计;

(2) 网上灯饰店数据库设计；

(3) 网上灯饰店主要模块的详细设计。

5.3.1 网上灯饰店功能结构设计

按照系统设计的原则,依据顶层数据流程图和第一层数据流图对系统的总体结构进行设计,系统总体结构包括用户信息管理、商品信息管理和订单信息管理三大模块。另外,考虑到系统的安全问题,需要在系统中增设数据备份等功能,在系统中增加了系统管理模块,系统总体功能结构如图 5.9 所示。

依据系统设计原则和第二层数据流图(用户信息管理),对用户信息管理模块进行进一步的分解设计,得到用户信息管理模块的功能结构如图 5.10 所示。

图 5.9 灯饰店总体功能结构图

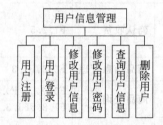
图 5.10 用户信息管理功能结构图

依据系统设计原则和第二层数据流图(商品信息管理),对商品信息管理模块进行进一步的分解设计,得到商品信息管理模块的功能结构如图 5.11 所示。

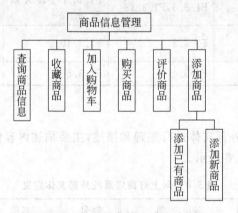
图 5.11 商品信息管理功能结构图

依据系统设计原则和第二层数据流图(订单信息管理),对订单信息管理模块进行进一步的分解设计,得到订单信息管理模块的功能结构如图 5.12 所示。

依据系统设计原则,对系统管理模块进行分解,得到系统管理模块的功能结构如图 5.13 所示。

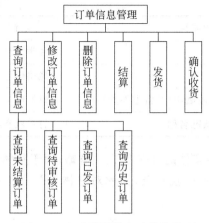

图 5.12 订单信息管理功能结构图

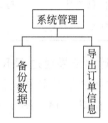

图 5.13 系统管理功能结构图

5.3.2 网上灯饰店数据库设计

在本部分设计中,主要进行的是概念设计和逻辑设计。

1. 数据库概念设计

依据数据分析,抽象、整理出的六项基本实体,为了使数据更规范增设了七项实体,基本实体的详细说明如表 5.9 所示,增设的实体详细说明如表 5.10 所示。

表 5.9 灯饰店中的基本实体

序号	实体名称	具有的属性	关键字
1	用户	用户编号、姓名、地址、邮编、电话号码、邮箱、性别、角色编号、状态编号、登录密码、最后登录时间	用户编号
2	订单	订单编号、购买件数、总价、下订单日期、订单状态编号、是否显示、收件人姓名、收件人电话、地址、邮箱、邮编、快递类型、支付种类、快递费用	订单编号
3	灯饰(商品)	灯饰编号、材质、颜色、安放位置、单价、库存量、商品详情、图片地址	灯饰编号
4	商品评价	评价编号、评价详情、评价日期	评价编号
5	收藏夹	收藏夹编号、收藏日期	收藏夹编号
6	购物车	购物车编号、添加日期	购物车编号

表 5.10 灯饰店中的增设实体

序号	实体名称	具有的属性	关键字
1	角色	角色编号、角色名称	角色编号
2	用户状态	用户状态编号、状态名称	用户状态编号
3	灯饰品牌	品牌编号、品牌名称	品牌编号

续表

序号	实体名称	具有的属性	关键字
4	灯饰种类	种类编号、种类名称	种类编号
5	灯饰风格	风格编号、风格名称	风格编号
6	灯饰信息统计	灯饰编号、点击次数、收藏次数、售出件数	灯饰编号
7	订单状态	订单状态编号,订单状态名称	订单状态编号

概念设计的结果通过 E-R 图描述出来,各实体的 E-R 图在此省略;由于整体概念模型(即实体之间联系 E-R 图)用一张图描述太大并且复杂,为此对整体 E-R 图进行了分解,分别表示在图 5.14～图 5.17 中。

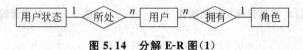

图 5.14　分解 E-R 图(1)

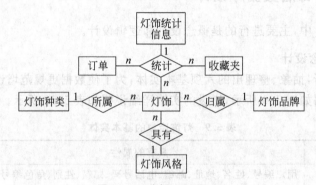

图 5.15　分解 E-R 图(2)

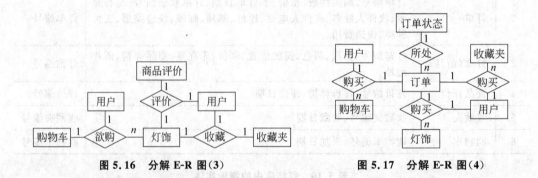

图 5.16　分解 E-R 图(3)　　　　图 5.17　分解 E-R 图(4)

2. 数据库逻辑设计

逻辑设计是将现实世界的概念模型设计成数据库的一种逻辑模式,即适应于某种特定数据库管理系统所支持的逻辑数据模式,因为 SQL Server 2008 是关系型数据库,因而按照规范化理论把 E-R 模型中的实体和联系转换为关系,进而设计出各种表。灯饰店系统中的关系如表 5.11 所示。

表 5.11 灯饰店系统中的关系

序号	实体名称	具有的属性	关键字
1	用户	用户编号、姓名、地址、邮编、电话号码、邮箱、性别、角色编号、状态编号、登录密码、最后登录时间	用户编号
2	订单	订单单编号、用户编号、灯饰编号、购买件数、总价、下订单日期、订单状态编号、是否显示、收件人姓名、收件人电话、地址、邮箱、邮编、快递类型、支付种类、快递费用	订单编号
3	灯饰(商品)	灯饰编号、种类编号、品牌编号、风格编号、材质、颜色、安放位置、单价、库存量、商品详情、图片地址	灯饰编号
4	商品评价	评价编号、灯饰编号、用户编号、评价详情、评价日期	评价编号
5	收藏夹	收藏夹编号、灯饰编号、用户编号、收藏日期	收藏夹编号
6	购物车	购物车编号、灯饰编号、用户编号、添加日期	购物车编号
7	角色	角色编号、角色名称	角色编号
8	用户状态	用户状态编号、状态名称	用户状态编号
9	灯饰品牌	品牌编号、品牌名称	品牌编号
10	灯饰种类	种类编号、种类名称	种类编号
11	灯饰风格	风格编号、风格名称	风格编号
12	灯饰信息统计	灯饰编号、点击次数、收藏次数、售出件数	灯饰编号
13	订单状态	订单状态编号,订单状态名称	订单状态编号

每种关系在 SQL Server 2008 对应的各表如表 5.12 至表 5.24 所示。

表 5.12 User(用户信息表)

序号	英文名	中文名	类型	长度/字节
1	UserId	用户编号	int	4
2	UserName	用户姓名	Varchar	20
3	Address	地址	Varchar	50
4	PostalCode	邮编	Varchar	6
5	Phone	电话号码	Varchar	11
6	Email	邮箱	Varchar	50
7	Gender	性别	Varchar	4
8	UserRoleId	用户角色编号	int	4
9	UserStatesId	用户状态编号	int	4
10	LoginId	登录用户名	Varchar	50
11	LoginPwd	登录密码	Varchar	50
12	LastLoginTime	最后登录时间	Varchar	10

表 5.13 UserRoles（角色信息表）

序号	英文名	中文名	类型	长度/字节
1	Id	角色编号	int	4
2	Name	角色名称	Varchar	20

表 5.14 UserStates（用户状态信息表）

序号	英文名	中文名	类型	长度/字节
1	Id	用户状态编号	int	4
2	Name	状态名称	Varchar	20

表 5.15 Brand（灯饰品牌表）

序号	英文名	中文名	类型	长度/字节
1	BrandId	品牌编号	int	4
2	BrandName	品牌名称	Varchar	20

表 5.16 Catagory（灯饰种类信息表）

序号	英文名	中文名	类型	长度/字节
1	CatagoryId	种类编号	int	4
2	CatagoryName	种类名称	Varchar	20

表 5.17 Style（灯饰风格信息表）

序号	英文名	中文名	类型	长度/字节
1	StyleId	风格编号	int	4
2	StyleName	风格名称	Varchar	10

表 5.18 Comments（商品评价信息表）

序号	英文名	中文名	类型	长度/字节
1	CommentsId	评价编号	int	4
2	LightId	灯饰编号	int	4
3	UserId	用户编号	int	4
4	Comments	评价详情	Varchar	150
5	Date	评价日期	Varchar	10

表 5.19 WishList(收藏夹信息表)

序号	英文名	中文名	类型	长度/字节
1	WishListId	收藏夹商品编号	int	4
2	LightId	灯饰编号	int	4
3	UserId	用户编号	int	4
4	StoreDate	收藏日期	Varchar	10

表 5.20 OrderStates(订单状态信息表)

序号	英文名	中文名	类型	长度/字节
1	Id	订单状态编号	int	4
2	Name	订单状态名称	Varchar	20

表 5.21 Light(灯饰信息表)

序号	英文名	中文名	类型	长度/字节
1	LightId	灯饰编号	int	4
2	CatagoryId	种类编号	int	4
3	BrandId	品牌编号	int	4
4	StyleId	风格编号	int	4
5	Stuff	材质	Varchar	20
6	Color	颜色	Varchar	20
7	Locate	安放位置	Varchar	20
8	Price	单价	decimal	18
9	Quantity	库存量	int	4
10	Description	商品详情	Varchar	250
11	Images	图片地址	Varchar	50

表 5.22 MyShopCart(购物车信息表)

序号	英文名	中文名	类型	长度/字节
1	ShopCartId	购物车商品编号	int	4
2	LightId	灯饰编号	int	4
3	UserId	用户编号	int	4
4	AddDate	添加日期	Varchar	10

表 5.23 Statistics(灯饰信息统计表)

序号	英文名	中文名	类型	长度/字节
1	LightId	灯饰编号	int	4
2	Clicks	点击次数	int	4
3	StoreCount	收藏次数	int	4
4	SaleCount	售出件数	int	4

表 5.24 Orders（订单信息表）

序号	英文名	中文名	类型	长度/字节
1	OrderId	订单编号	Varchar	17
2	UserId	用户编号	int	4
3	LightId	灯饰编号	int	4
4	Quantity	购买数量	int	4
5	TotalPrice	总价	decimal	18
6	Date	下订单日期	Varchar	10
7	OrderStateId	订单状态编号	int	4
8	ShowOrNot	是否显示	int	4
9	ReceiveName	收件人姓名	Varchar	20
10	ReceivePhone	收件人电话	Varchar	11
11	Address	地址	Varchar	50
12	Email	邮箱	Varchar	50
13	PostalCode	邮编	Varchar	6
14	PostType	快递类型	Varchar	20
15	PaymentType	支付类型	Varchar	20
16	PostFee	快递费用	int	4

5.3.3 主要模块功能详细设计

1. 用户信息管理模块功能详细设计

用户信息管理功能主要是对用户的个人信息进行管理，包括新用户注册成功时将用户信息录入数据库系统，已注册用户查看自己的用户信息或根据自己的实际情况修改个人资料，如果觉得自己的密码不够安全，可以修改登录密码。而管理员则可以查看已有三个月或更久时间未登录系统的用户基本信息，有必要的话，管理员有权将这些用户删除。其程序流程图如图 5.18 所示。

2. 商品信息管理模块功能详细设计

商品信息管理功能主要是对系统中的灯饰信息进行管理，包括对灯饰信息的查询、按条件搜索符合条件的灯饰信息、将灯饰加入收藏夹、将灯饰添加到购物车、购买商品、评价商品及管理员添加新品的相关功能的管理。用户登录与否都可以查看商品信息或搜索符合条件的商品信息，但是除此之外的其他操作都需要用户先登录系统才能进行。用户收藏夹和购物车中商品的信息只要是用户不删除，系统就会一直记录在相应的数据表中，用户登录就可以查看到自己的收藏记录和购物车添加的商品信息。

管理员拥有普通用户可以进行操作的所有与商品信息相关的功能权限，并且添加新

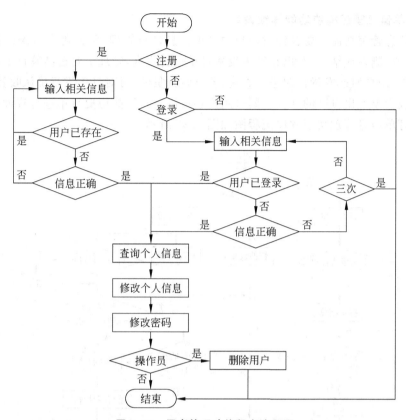

图 5.18 用户管理功能程序流程图

品功能是系统管理员才具有的权限,只有管理员才能对该功能进行操作。商品信息管理功能的程序流程图如图 5.19 所示。

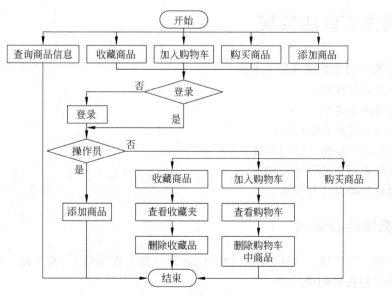

图 5.19 商品信息管理功能程序流程图

3. 订单信息管理模块功能详细设计

订单信息管理功能主要是针对用户的订单信息进行管理，包括用户对未结算订单进行查询、修改、删除或结算，对已结算未发货订单进行查询，管理员对已结算订单进行查询及发货操作，用户对已发货订单进行查询，若已收到货物，可对订单进行确认收货操作，还可对商品或卖家做出自己的评价。同时用户还可以对自己的历史订单进行查询或清理操作。其中订单信息管理功能程序流程图如图 5.20 所示。

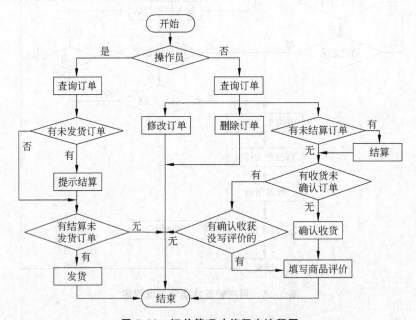

图 5.20　订单管理功能程序流程图

5.4　网上灯饰店实现

在设计和实现主要完成如下工作：
(1) 系统运行环境；
(2) 数据库的建立；
(3) 系统实现的总体框架；
(4) 数据访问层的设计与实现；
(5) 业务逻辑层的设计与实现；
(6) 表示层的设计与实现。

5.4.1　系统运行环境

网络环境：在已建立的局域网及全国广域网框架上均可运行，基于以 TCP/IP 传输协议为基础的数据联网模式。

后台数据库：SQL Server 2008。

开发软件平台：Visual Studio 2010（ASP.NET）。

服务器操作系统平台：Windows 2007 Serrer/Windows 2003 1GB 内存。

客户机：Windows 2007 Professional/Windows 2007 Server/Windows XP/Windows 2007,1024×768 分辨率及以上,256MB 以上内存。

5.4.2 数据库的建立与连接

网上灯饰店系统中数据库命名为 MyLightShop,在数据库设计中设计的各种表在数据库中实现,同时建立一些必要的视图以方便查询,在数据库中建立的各种表和视图如图 5.21 所示。

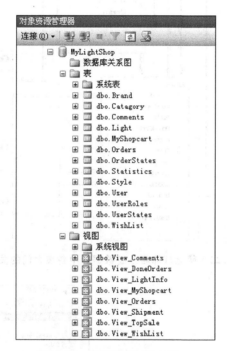

图 5.21 数据库 MyLightShop

各种表之间的关系如图 5.22 所示。

5.4.3 系统实现的总体框架

系统实现的总体架构如图 5.23 所示,在总体架构中包含五个部分,即数据访问层 DAL、业务逻辑层 BLL、接口 Model、公共库 CommonLibrary、表示层和用户应用程序 MyLightShop。

5.4.4 数据访问层的设计与实现

在数据访问层中设计实现了 11 个数据访问类,如图 5.24 所示。

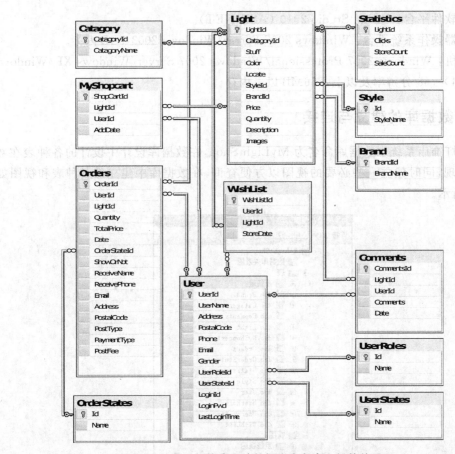

图 5.22 网上灯饰店系统数据库中各表之间的关系

图 5.23 系统总体框架

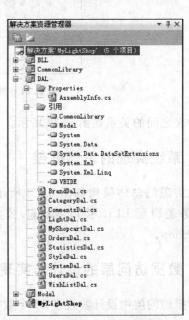

图 5.24 数据访问层的构成

1. BrandDal

该类的主要数据访问操作有：

（1）DataTable GetBrand()，查询所有灯饰品牌，在 Br0and 表上操作；

（2）DataTable GetBrandById(string brandId)，根据品牌编号查询灯饰品牌，在 Brand 表上操作；

（3）DataTable IsExist(string brandName)，判断某种灯饰品牌是否存在，在 Brand 表上操作；

（4）bool InsertIntoBrand(string name)，添加某种灯饰品牌，在 Br0and 表上操作。

2. CategoryDal

该类的主要数据访问操作有：

（1）DataTable GetCategory()，查询所有灯饰种类，在 Catagory 表上操作；

（2）DataTable GetCategoryById(string categoryId)，根据种类编号查询灯饰种类，在 Catagory 表上操作；

（3）DataTable IsExist(string categoryName)，根据种类名称判断某种灯饰种类是否存在，在 Catagory 表上操作；

（4）bool InsertIntoCategory(string name)，添加某种灯饰种类，在 Catagory 表上操作。

3. CommentsDal

该类的主要数据访问操作有：

（1）DataTable GetCommentsByLightId(int lightId)，通过灯饰编号查询商品评价，在 View_Comments 视图上操作；

（2）bool InsertComment(Comments comment)，添加对某一灯饰的评价，在 Comments 表上操作。

4. LigthDal

该类的主要数据访问操作有：

（1）DataTable GetLightByLightId(string lightId)，通过灯饰编号查询商品的详细信息，在 View_LightInfo 视图上操作；

（2）bool DeleteLightById(string lightId)，通过灯饰编号删除某种灯饰，在 Light 表上操作；

（3）bool UpdateLightById(Light light)，通过灯饰编号更新某种灯饰商品，在 Light 表上操作；

（4）DataTable GetLights()，查询所有灯饰商品的信息，在 Light 表上操作；

（5）DataTable GetAllLightById(string lightId)，查询灯饰商品所有的详细信息，在 View_LightInfo 视图上操作；

（6）DataTable GetAllLightById(string lightId，通过灯饰编号查询商品的详细信息，在 View_LightInfo 视图上操作；

（7）DataTable GetLightByCatagory(string catagoryId)，通过种类编号查询商品的

详细信息,在 View_LightInfo 视图上操作;

(8) DataTable GetLightsById(string lightId),通过灯饰编号查询商品的详细信息,在 View_LightInfo 视图上操作;

(9) bool InsertLightInfo(Light light),添加新的灯饰商品,对 Light 表操作;

(10) DataTable GetLightByInfo(string sql):通过给定的查询条件查询商品的信息,在 Light 表上操作;

(11) DataTable GetLightsByBrandId(string brandId),通过品牌编号查询商品的详细信息,在 View_LightInfo 视图上操作;

(12) DataTable GetLightsByStyleId(string styleId),通过风格编号查询商品的详细信息,在 View_LightInfo 视图上操作;

(13) bool AddQuantity(string lightId,int quantity),修改某种灯饰的数量,在 Light 表上操作;

(14) bool ShipLight(int count),修改所有灯饰商品的数量为给定数量,在 Light 表上操作。

5. MyShopcartDal

该类的主要数据访问操作有:

(1) DataTable GetMyShopCartByUserId(string userId),查询某一用户的购物车信息,在 View_MyShopcart 视图上操作;

(2) bool InsertIntoMyShopcart(string userId, string lightId,string date),向某一用户的购物车中添加某种灯饰商品,在 MyShopcart 表上操作;

(3) bool DeleteShopcartByLightId(string userId, string lightId),删除某一用户的购物车中的某种灯饰商品,在 MyShopcart 表上操作;

(4) bool DeleteShopcartByUserId(string userId),清空某一用户的购物车,在 MyShopcart 表上操作;

(5) DataTable IsExist(string lightId, string userId),判断某一用户的购物车中是否有某种灯饰商品,在 MyShopcart 表上操作。

6. OrdersDal

该类的主要数据访问操作有:

(1) DataTable GetOrdersByLightId(string lightId),通过灯饰编号查询已完成的订单信息,在 View_Orders 视图上操作;

(2) DataTable GetWaitHandleOrderByUserId(string userId),通过用户编号查询该用户存在的已发货的订单,在 View_Orders 视图上操作;

(3) DataTable GetNotHandleOrderByUserId(string userId),通过用户编号查询该用户存在的待审查的订单,在 View_Orders 视图上操作;

(4) DataTable GetWaitShipOrderByUserId(string userId),通过用户编号查询该用户存在的正在配货的订单,在 View_Orders 视图上操作;

(5) DataTable GetHandledOrderByUserId(string userId),通过用户编号查询该用

户存在的已完成并且可显示的订单,在 View_Orders 视图上操作;

（6）bool InsertIntoOrders(Orders order),添加订单,对 Orders 表操作;

（7）DataTable GetOrdersByOrderId(string orderId),按订单编号查询订单,在 View_Orders 视图上操作;

（8）bool DeleteOrdersByOrderId(string orderId),按订单编号删除订单,在 Orders 表上操作;

（9）bool DeleteHandledOrdersByUserId(string userId),按订单编号修改订单为不显示,在 Orders 表上操作;

（10）bool DeleteHandledOrdersByOrderId(string orderId),把某一用户已完成的订单修改为不可显示,在 Orders 表上操作;

（11）bool AffirmReceived(string orderId),把某一订单的状态改为已完成,在 Orders 表上操作;

（12）bool CheckOut(Orders order),配置用户订单中的收货人信息,并且把状态设置为正在配货,在 Orders 表上进行操作;

（13）DataTable GetShipmentOrders(),把某一订单收货人的相关信息,在 View_Orders 视图上操作;

（14）bool ShipLight(string orderId),把某一订单的状态改为已发货,在 Orders 表上操作;

（15）bool UpdateOrder(Orders order),把已发货的订数量和金额累加订单总量和总金额中,在 Orders 表上操作。

7. StatisticsDal

该类的主要数据访问操作有：

（1）DataTable GetTopSale(),查询销售数量最多的五种灯饰商品,在 View_TopSale 视图上操作;

（2）bool AddClicks(string lightId),点击计数,在 Statistics 表上操作;

（3）bool AddStoreCount(string lightId),修改某种销售数量,在 Statistics 表上操作;

（4）bool AddSaleCount(string lightId,string salecount),把某种销售数量累计到总销售数量中,在 Statistics 表上操作。

8. StyleDal

该类的主要数据访问操作有：

（1）DataTable GetStyle(),查询所有的灯饰风格,在 Style 表上操作;

（2）DataTable GetStyleById(string styleId),按灯饰风格编号查询灯饰风格,在 Style 表上操作;

（3）DataTable IsExist(string styleName),按灯饰风格编号判断某种风格灯饰是否存在,在 Style 表上操作;

（4）bool InsertIntoStyle(string name),添加某种灯饰风格,在 Style 表上操作。

9. WishListDal

该类的主要数据访问操作有:

(1) DataTable GetMyStoreByUserId(string userId),获得某一用户的收藏信息,在 View_WishList 视图上操作;

(2) bool InsertIntoWishList(string userId, string lightId, string date),某一用户向收藏夹中加入某一商品,在 WishList 表上操作;

(3) bool DeleteWishListByLightId(string userId, string lightId),某一用户删除收藏夹中某一商品,在 WishList 表上操作;

(4) bool DeleteWishListByUserId(string userId),删除某一用户收藏夹中所有商品,在 WishList 表上操作;

(5) DataTable IsExist(string lightId, string userId),判断某一用户是否收藏了某一商品,在 WishList 表上操作。

10. SystemDal

该类的主要数据访问操作有:

(1) 备份数据库到指定存储器中实现的代码如下。

```
public bool BackUp(string path)
    {
string sql="backup database MyLightShop to disk='"+path+"' with init;";
try
        {
            SqlCommand sqlCommand=DBClass.GetCommand(sql);
flag=DBClass.ExecNonQuery(sqlCommand);
        }
catch(Exception)
        {
            //
            //TODO:
            //
        }
return flag;
        }
```

(2) public DataTable GetReportView(string date),查询某一日期成交的订单,在对 View_DoneOrders 视图上操作。

11. UsersDal

该类的主要数据访问操作有:

(1) 用户登录;

(2) 用户注册;

(3) 修改用户信息;

(4) 通过登录名称判断用户是否存在；
(5) 通过用户编号查询用户信息；
(6) 查询所有用户信息；
(7) 删除用户；
(8) 更新用户最后登录时间；
(9) 修改用户密码。

在 Visual Studio 2010 中的实现代码在如下：

```
namespace DAL
{
public class UsersDal
    {
        DataTable dataTable=null;
        public DataTable UserLogin(string loginId, string loginPwd,int roleId)
          //用户登录
        {
string sql="select * from [User] where LoginId='"+loginId+"' and LoginPwd='"+loginPwd+"' and UserRoleId='"+roleId+"';";
try
        {
            SqlCommand sqlCommand=DBClass.GetCommand(sql);
DBClass.ExecNonQuery(sqlCommand);
dataTable=DBClass.GetDataSet(sqlCommand, "User");
        }
catch(Exception)
        {
            //
            //TODO:
            //
        }
return dataTable;
        }
        public bool Resgister(User user)             //用户注册
        {
bool flag=false;
            string sql="insert into [User] ([UserName],[Address],[PostalCode],[Phone],[Email],[Gender],[UserRoleId],[UserStateId],[LoginId],[LoginPwd],[LastLoginTime]) values ('"+user.userName+"','"+user.address+"','"+user.postalCode+"','"+user.phone+"','"+user.email+"','"+user.gender+"',2,1,'"+user.loginId+"','"+user.loginPwd+"','"+user.lastLoginTime+"');";
try
        {
```

```csharp
                    SqlCommand sqlCommand=DBClass.GetCommand(sql);
    flag=DBClass.ExecNonQuery(sqlCommand);
                }
    catch(Exception)
                {
                    //
                    //TODO:
                    //
                }
    return flag;
            }
            public bool Modify(User user)                    //修改用户信息
            {
bool flag=false;
                string sql=" update [User] set [UserName] = '" + user.userName +"',
                [Address]='"+user.address+"',[PostalCode]='"+user.postalCode+"',
                [Phone] = '" + user. phone +"', [Email] = '" + user. email +"'    where
                [LoginId]='"+user.loginId+"';";
    try
                {
                    SqlCommand sqlCommand=DBClass.GetCommand(sql);
    flag=DBClass.ExecNonQuery(sqlCommand);
                }
    catch(Exception)
                {
                    //
                    //TODO:
                    //
                }
    return flag;
            }
            public DataTable IsExists(string loginId)      //通过登录名称判断用户是否存在
            {
                string sql="select * from [User] where LoginId='"+loginId+"'";
    try
                {
                    SqlCommand sqlCommand=DBClass.GetCommand(sql);
DBClass.ExecNonQuery(sqlCommand);
dataTable=DBClass.GetDataSet(sqlCommand, "User");
                }
    catch(Exception)
                {
                    //
                    //TODO:
```

```csharp
            //
        }
        return dataTable;
    }
    public DataTable GetUserInfoByUserId(string userId)          //通过用户编号查询用户信息
    {
        string sql="select * from [User] where [UserId]='"+int.Parse(userId)+"';";
        try
        {
            SqlCommand sqlCommand=DBClass.GetCommand(sql);
            DBClass.ExecNonQuery(sqlCommand);
            dataTable=DBClass.GetDataSet(sqlCommand, "UserInfo");
        }
        catch(Exception)
        {
            //
            //TODO:
            //
        }
        return dataTable;
    }
    public DataTable GetAllUserInfo(string date)     //查询所有用户信息
    {
        string sql="select * from [User] where [LastLoginTime] like '"+date+"';";
        try
        {
            SqlCommand sqlCommand=DBClass.GetCommand(sql);
            DBClass.ExecNonQuery(sqlCommand);
            dataTable=DBClass.GetDataSet(sqlCommand, "UserInfo");
        }
        catch(Exception)
        {
            //
            //TODO:
            //
        }
        return dataTable;
    }
    public bool DeleteUserByUserId(string userId)   //删除用户
    {
        string sql="delete from [User] where [UserId]='"+int.Parse(userId)+"';";
        bool flag=false;
        try
```

```csharp
            {
                SqlCommand sqlCommand=DBClass.GetCommand(sql);
    flag=DBClass.ExecNonQuery(sqlCommand);
            }
    catch(Exception)
            {
                //
                //TODO:
                //
            }
    return flag;
        }
        public bool UpdateLastLoginTime(string userId,string date)         //更新用户最后登录时间
            {
string sql="update [User] set [LastLoginTime]='"+date+"' where [UserId]='"+int.Parse(userId)+"';";
bool flag=false;
try
            {
                SqlCommand sqlCommand=DBClass.GetCommand(sql);
    flag=DBClass.ExecNonQuery(sqlCommand);
            }
    catch(Exception)
            {
                //
                //TODO:
                //
            }
    return flag;
        }
        public bool UpdatePwd(int userId, string pwd)       //修改用户密码
            {
string sql="update [User] set [LoginPwd]='"+pwd+"' where [UserId]='"+userId+"';";
bool flag=false;
try
            {
                SqlCommand sqlCommand=DBClass.GetCommand(sql);
    flag=DBClass.ExecNonQuery(sqlCommand);
            }
    catch(Exception)
            {
                //
                //TODO:
```

```
                //
          }
return flag;
        }
    }
}
```

5.4.5 业务逻辑层的设计与实现

在业务逻辑层设计实现了与数据访问层中的 11 个数据访问类相对应的 11 个业务逻辑类,如图 5.25 所示。

（1）BrandBLL：针对品牌业务逻辑设计实现的类。

（2）CategoryBLL：针对灯饰种类业务逻辑设计实现的类。

（3）CommentsBLL：针对商品评价业务逻辑设计实现的类。

（4）LigthBLL：针对灯饰信息业务逻辑设计实现的类。

（5）MyShopCartBLL：针对购物车业务逻辑设计实现的类。

图 5.25 业务逻辑层的构成

（6）OrdersBLL：针对订单信息业务逻辑设计实现的类。

（7）StatisticsBLL：针对灯饰统计业务逻辑设计实现的类。

（8）StyleBLL：针对灯饰风格业务逻辑设计实现的类。

（9）WishListBLL：针对收藏夹业务逻辑设计实现的类。

（10）SystemBLL：针对系统管理业务逻辑设计实现的类。

（11）UsersBLL：针对品牌业务逻辑设计实现的类；具体实现代码如下：

```
namespace BLL
{
public class UsersBll
    {
        UsersDal userDal=new UsersDal();
        DataTable dataTable=null;
public DataTable UserLogin(string loginId, string loginPwd,int roleId)
                                                    //用户登录
        {
try
        {
dataTable=userDal.UserLogin(loginId, loginPwd,roleId);
        }
```

```
            catch(Exception)
                {
                    //
                    //TODO:
                    //
                }
            return dataTable;
        }
        public DataTable IsExist(string loginId)          //判断用户是否存在
        {
            try
                {
                    dataTable=userDal.IsExists(loginId);
                }
            catch(Exception)
                {
                    //
                    //TODO:
                    //
                }
            return dataTable;
        }
        public bool Resgister(User user)                   //用户注册
        {
            bool flag=false;
            try
                {
                    flag=userDal.Resgister(user);
                }
            catch(Exception)
                {
                    //
                    //TODO:
                    //
                }
            return flag;
        }
        public bool Modify(User user)                      //修改用户信息
        {
            bool flag=false;
            try
                {
                    flag=userDal.Modify(user);
                }
            catch(Exception)
                {
```

```
                //
                //TODO:
                //
            }
return flag;
        }
public DataTable GetUserInfoByUserId(string userId)    //根据用户编号获取用户信息
        {
try
            {
dataTable=userDal.GetUserInfoByUserId(userId);
            }
catch(Exception)
            {
                //
                //TODO:
                //
            }
return dataTable;
        }
public DataTable GetAllUserInfo(string date)    //根据日期获取用户信息
        {
try
            {
dataTable=userDal.GetAllUserInfo(date);
            }
catch(Exception)
            {
                //
                //TODO:
                //
            }
return dataTable;
        }
public bool DeleteUserByUserId(string userId)    //根据用户编号删除用户
        {
bool flag=false;
try
            {
flag=userDal.DeleteUserByUserId(userId);
            }
catch(Exception)
            {
                //
                //TODO:
                //
```

```
            }
    return flag;
         }
public bool UpdateLastLoginTime(string userId, string date)
                                                    //更改指定用户的最后登录时间
         {
bool flag=false;
try
         {
flag=userDal.UpdateLastLoginTime(userId,date);
         }
catch(Exception)
         {
             //
             //TODO:
             //
         }
    return flag;
         }
public bool UpdatePwd(int userId, string pwd)    //修改指定用户的密码
         {
bool flag=false;
try
         {
flag=userDal.UpdatePwd(userId, pwd);
         }
catch(Exception)
         {
             //
             //TODO:
             //
         }
    return flag;
         }
         }
}
```

5.4.6 公共库的设计与实现

设计公共库的目的是提高系统的健壮性、复用性，公共库构成如图5.26所示。

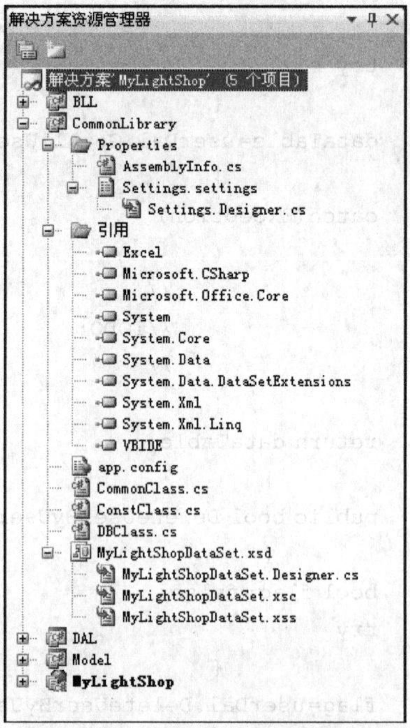

图 5.26 公共库的构成

1. 应用配置 app.config，设置数据库源

相应地，实现代码如下：

```
<configuration>
<configSections>
</configSections>
<connectionStrings>
<add name="CommonLibrary.Properties.Settings.MyLightShopConnectionString"
connectionString=" Data Source = 5QNTOGIN9V7GNT8; Initial Catalog = MyLightShop;
Integrated Security=True"
providerName="System.Data.SqlClient" />
</connectionStrings>
</configuration>
```

2. 数据库连接和操作 DBClass.cs

相应地,实现代码如下:

```
namespace CommonLibrary
{
public class DBClass
    {
public static SqlConnection GetConnection()
        {
string strConnection =" Data Source = 5QNTOGIN9V7GNT8.; Initial Catalog = MyLightShop;Integrated Security=True";
            //建立连接
            SqlConnection sqlConnection=new SqlConnection(strConnection);
return sqlConnection;
        }
public static SqlCommand GetCommand(string sql)
        {
            SqlCommand command=new SqlCommand(sql, GetConnection());
return command;
        }
public static bool ExecNonQuery(SqlCommand sqlCommand)
        {
bool flag=false;
try
            {
if(sqlCommand.Connection.State !=ConnectionState.Open)
                {
sqlCommand.Connection.Open();
                }
flag=sqlCommand.ExecuteNonQuery()>0;
            }
catch(Exception exception)
                {
```

```csharp
            throw new Exception(exception.Message, exception);
            //
            //TODO:
            //
        }
        finally
        {
            if(sqlCommand.Connection.State==ConnectionState.Open)
            {
                sqlCommand.Connection.Close();
            }
        }
        return flag;
    }
    public static DataTable GetDataSet(SqlCommand sqlCommand, string tableName)
    {
        SqlDataAdapter sqlDataAdapter;
        DataSet dataSet=new DataSet();
        try
        {
            if(sqlCommand.Connection.State !=ConnectionState.Open)
            {
                sqlCommand.Connection.Open();
            }
            sqlDataAdapter=new SqlDataAdapter(sqlCommand);
            sqlDataAdapter.Fill(dataSet, tableName);
            return dataSet.Tables[tableName];
        }
        catch(Exception exception)
        {
            throw new Exception(exception.Message, exception);
            //
            //TODO:
            //
        }
        finally
        {
            if(sqlCommand.Connection.State==ConnectionState.Open)
            {
                sqlCommand.Connection.Close();
            }
        }
    }
}
```

3．ConstClass.cs：常量定义

具体内容如下：

```
namespace CommonLibrary
{
public class ConstClass
    {
public static string USERFORADMAIN="Admin";
public static string USERFORMEMBER="Member";
public static int QUANTITY_FALSE=0;
public static int QUANTITY_TRUE=1;
    }
}
```

4．CommonClass.cs 公共对话框的处理

具体代码如下：

```
namespace CommonLibrary
{
public class CommonClass
    {
        ///在客户端弹出对话框
public static string MessageBox(string txtMessage)
        {
string str;
str="<script language=javascript>alert('"+txtMessage+"')</script>";
return str;
        }
        ///在客户端弹出对话框,关闭对话框返回指定页
public static string MessageBox(string txtMessage, string url)
        {
string str;
str="<script language=javascript>alert('"+txtMessage+"');location='"+url+"';</script>";
return str;
        }
public static string MessageBoxPage(string txtMessage)
        {
string str;
str =" < script language = javascript > alert ('" + txtMessage +"'); location =
'javascript:history.go(-1)';</script>";
return str;
        }

public string RandomNum(int n)
        {
            string strChar="0,1,2,3,4,5,6,7,8,9,A,B,C,D,E,F,G,H,I,J,K,L,M,N,O,
```

P,Q,R,S,T,U,V,W,X,Y,Z,a,b,c,d,e,f,g,h,i,j,k,l,m,n,o,p,q,r,s,t,u,v,w,x,y,z";

```
string[] array=strChar.Split(',');
string num="";
int temp=-1;
            Random random=new Random();
for(int i=1; i<n+1; i++)
            {
if(temp !=-1)
                {
random=new Random(i * temp * unchecked((int)DateTime.Now.Ticks));
                }
int t=random.Next(61);
if(temp !=-1 && temp==t)
                {
return RandomNum(n);
                }
temp=t;
num+=array[t];
                }
return num;
        }
    }
}
```

5.4.7 实体模型部分的设计与实现

设计实体模型部分的目的是提高系统的封装性、统一性和复用性,实体模型部分构成如图 5.27 所示。

针对各个表设计了相应的实体模型,表中的字段对应实体模型的属性,针对各属性设置了设置器和访问器。

1. 针对品牌表 brand 设置的实体模型

brand 实体模型中的属性有：

public int brandId
public string brandName

2. 针对灯饰种类表 catagory 设置的实体模型

catagory 实体模型中的属性有：

public int catagoryId
public string catagoryName

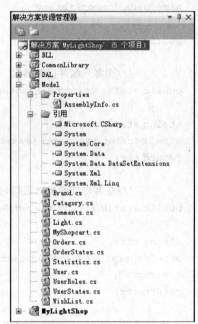

图 5.27 接口部分的构成

3. 针对商品评价表 comments 设置的实体模型
comments 实体模型中的属性有：

```
public int commentsId
public int userId
public int lightId
public string comments
public string  date
```

4. 针对灯饰信息表 light 设置的实体模型
light 实体模型的属性有：

```
public int lightId
public int catagoryId
public string stuff
public string color
public string locate
public int styleId
public int brandId
public decimal price
public int clicks
public int quantity
public string description
public string images
```

5. 针对购物车表 shopcart 设置的实体模型
shopcart 实体模型的属性有：

```
public int shopcartId
public int lightId
```

6. 针对订单信息表 orders 设置的实体模型
orders 实体模型的属性有：

```
public string ordersId
public string orderDate
public int userId
public int lightId
public int quantity
public decimal totalPrice
public int orderStateId
public int showOrNot
public string receiveName
public string receivePhone
public string email
public string address
public string postalCode
```

```
public string postType
public string paymentType
public int postFee
```

7. 针对订单状态表 orderStates 设置的实体模型
orderStates 实体模型的属性有：

```
public int orderStatesId
public string orderStatesName
```

8. 针对订单信息统计表 Statics 设置的实体模型
Statics 实体模型的属性有：

```
public int lightId
public int clicks
public int storeCount
public int saleCount
```

9. 针对用户角色表 userRoles 设置的实体模型
userRoles 实体模型的属性有：

```
public int userRolesId
public string userRolesName
```

10. 针对用户状态表 userStates 设置的实体模型
userStates 实体模型的属性有：

```
public int userStatesId
public string userStatesName
```

11. 针对收藏夹表 wishList 设置的实体模型
wishList 实体模型的属性有：

```
public int wishListId
public int userId
public int lightId
```

12. 针对用户信息表 User 设置的实体模型
User 实体模型实现如下：

```
namespace Model
{
public class User
    {
public int userId
        {
set;
```

```
            get;
        }
public string userName
        {
set;
get;
        }
public string address
        {
set;
get;
        }
public string postalCode
        {
set;
get;
        }
public string phone
        {
set;
get;
        }
public string email
        {
set;
get;
        }
public string gender
        {
set;
get;
        }
public UserRoles userRole
        {
set;
get;
        }
public UserStates userState
        {
set;
get;
        }
public string loginId
        {
set;
get;
        }
```

```
        public string loginPwd
            {
set;
get;
            }
        public string lastLoginTime
            {
set;
get;
            }
        }
    }
```

5.4.8 部分表示层及控制层的设计与实现

灯饰店的表示层及控制层的构成如图 5.28 所示。

图 5.28 表示层及控制层的构成

1. 系统主页的设计与实现

系统主页如图 5.29 所示，主页的整体结构为上下结构，整个页面分为三部分：

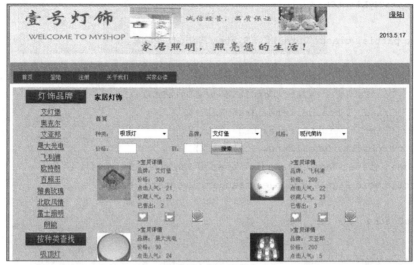

(a) 系统主页上半部分

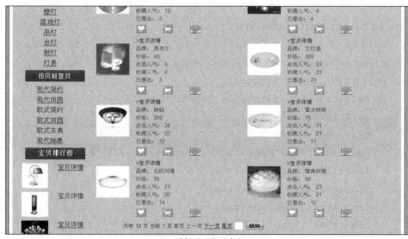

(b) 系统主页下半部分

图 5.29　系统主页

（1）页面顶部，用于展示网上商店的名称、标语等；

（2）页面左部，显示灯饰种类信息、灯饰风格信息、灯饰品牌信息及最畅销的商品信息，单击链接即可查看相关商品信息；

（3）页面的主要内容，默认为所有商品信息，可以根据用户需求，输入查询条件搜索商品信息，或者查看商品详情。主页面默认显示的是所有的商品信息，由于商品信息量较多，所以在页面底部设有分页功能，每页显示十件商品的信息。用户可以根据自己的需要进行收藏商品信息、将商品添加到购物车、查看商品详细信息、搜索符合查询条件的商品信息等操作。

下面是三部分的具体实现代码。

主页面右部实现代码：Default.aspx.cs

```csharp
namespace MyLightShop
{
    public partial class _Default : System.Web.UI.Page
    {
        DataTable dataTable=new DataTable();
        DataTable dataTableForClass=new DataTable();
        BrandBll brandBll=new BrandBll();
        LightBll lightBll=new LightBll();
        StyleBll styleBll=new StyleBll();
    protected static PagedDataSource ps=new PagedDataSource();
    protected void Page_Load(object sender, EventArgs e)
        {
    if(!IsPostBack)
            {
    DataListofGoodsBind();
            }
        }
    public void DataListofGoodsBind()
        {
    int currentPage=Convert.ToInt32(labNowPage.Text);
    dataTable=lightBll.GetAllLights();
            ps.DataSource=dataTable.DefaultView;
            ps.AllowPaging=true;
            ps.PageSize=10;
            ps.CurrentPageIndex=currentPage-1;        //取得当前页的页码
            lbkbtnFront.Enabled=true;
            lbkbtnFirst.Enabled=true;
            lbkbtnNext.Enabled=true;
            lbkbtnLast.Enabled=true;
    if(currentPage==1)
            {
                lbkbtnFirst.Enabled=false;            //不显示第一页按钮
                lbkbtnFront.Enabled=false;            //不显示上一页按钮
            }
    if(currentPage==ps.PageCount)
            {
                lbkbtnNext.Enabled=false;             //不显示下一页
                lbkbtnLast.Enabled=false;             //不显示最后一页
            }
            this.labCount.Text=Convert.ToString(ps.PageCount);
            DataList.DataSource=ps;
            DataList.DataKeyField="LightId";
    DataList.DataBind();
        }
```

```
protected void lbkbtnFirst_Click(object sender, EventArgs e)
        {
             this.labNowPage.Text="1";
DataListofGoodsBind();
        }
protected void lbkbtnFront_Click(object sender, EventArgs e)
        {
             this. labNowPage. Text = Convert. ToString (Convert. ToInt32 (this.
             labNowPage.Text)-1);
DataListofGoodsBind();
        }
protected void lbkbtnNext_Click(object sender, EventArgs e)
        {
             this. labNowPage. Text = Convert. ToString (Convert. ToInt32 (this.
             labNowPage.Text)+1);
DataListofGoodsBind();
        }
protected void lbkbtnLast_Click(object sender, EventArgs e)
        {
             this.labNowPage.Text=this.labCount.Text;
DataListofGoodsBind();
        }
protected void ChangeButton_Click(object sender, EventArgs e)
        {
int myPageNum=1;
if(!txtPage.Text.Equals(""))
            {
myPageNum=Convert.ToInt32(txtPage.Text.ToString());
            }
if(myPageNum<=0 || myPageNum>ps.PageCount)
                {
                    Response.Write(CommonClass.MessageBox("请输入页数并确定没有超过
                    总页数"));
                }
else
            {
                 this.labNowPage.Text=Convert.ToString(myPageNum);
DataListofGoodsBind();
            }
        }
protected void DataList_ItemCommand(object source, DataListCommandEventArgs e)
            {
if(e.CommandName=="detail")
            {
```

```csharp
                    StatisticsBll staBll=new StatisticsBll();
            if(staBll.AddClicks(e.CommandArgument.ToString()))
            Response.Redirect("~/DetailInfoDefault.aspx?LightId="+e.CommandArgument);
                    }
            if(e.CommandName=="buy")
                    {
                    Response.Write(CommonClass.MessageBox("您还未登录,请先登录!",
                    "Login.aspx?LightIdBuy="+e.CommandArgument));

                    }
            if(e.CommandName=="store")
                    {
                    Response.Write(CommonClass.MessageBox("您还未登录,请先登录",
                    "Login.aspx?LightIdStore="+e.CommandArgument));

                    }
            if(e.CommandName=="comment")
                    {
                    Response.Write(CommonClass.MessageBox("您还未登录,请先登录?",
                    "Login.aspx?LightIdComment="+e.CommandArgument));
                    }
                    }
protected void ButtonSearch_Click(object sender, EventArgs e)
                    {
string category=CategoryListBox.SelectedValue;
string style=StyleListBox.SelectedValue;
string brand=BrandListBox.SelectedValue;
string low=TextBoxLow.Text.Trim();
string high=TextBoxHigh.Text.Trim();
string sql="";
if(!low.Equals("")&& !high.Equals(""))
sql="select * from [View_LightInfo] where [CatagoryName]='"+category+"' and
[StyleName]='"+style+"' and [BrandName]='"+brand+"' and [Price] between '"+
int.Parse(low)+"' and '"+int.Parse(high)+"';";
else sql="select * from [View_LightInfo] where [CatagoryName]='"+category+"'
and [StyleName]='"+style+"' and [BrandName]='"+brand+"';";

Response.Redirect("~/LightListForDefault.aspx?Sql="+sql);
                    }
            }
    }
```

主页面左部对应的实现代码:Manage.Master.cs

```csharp
namespace MyLightShop
{
```

```csharp
public partial class Manager : System.Web.UI.MasterPage
    {
protected void Page_Load(object sender, EventArgs e)
        {
if(!IsPostBack)
            {
DataListofBind();
            }
        }
public void DataListofBind()
        {
            DataTable dataTableForBrand=new DataTable();
            BrandBll brandBll=new BrandBll();
dataTableForBrand=brandBll.GetBrand();
            DataList1.DataSource=dataTableForBrand.DefaultView;
            DataList1.DataKeyField=dataTableForBrand.Columns[0].ToString();
DataList1.DataBind();
            DataTable dataTableForStyle=new DataTable();
            StyleBll stylebll=new StyleBll();
dataTableForStyle=stylebll.GetStyle();
            DataList2.DataSource=dataTableForStyle.DefaultView;
            DataList2.DataKeyField=dataTableForStyle.Columns[0].ToString();
DataList2.DataBind();
            DataTable dataTableForTopSale=new DataTable();
            StatisticsBll stabll=new StatisticsBll();
dataTableForTopSale=stabll.GetTopSale();
            DataList3.DataSource=dataTableForTopSale.DefaultView;
            DataList3.DataKeyField=dataTableForTopSale.Columns[0].ToString();
DataList3.DataBind();
            DataTable dtForCategory=new DataTable();
            CategoryBll categoryBll=new CategoryBll();
dtForCategory=categoryBll.GetCategory();
            DataList4.DataSource=dtForCategory.DefaultView;
            DataList4.DataKeyField=dtForCategory.Columns[0].ToString();
DataList4.DataBind();
        }
protected void DataList1_ItemCommand(object source, DataListCommandEventArgs e)
        {
if(e.CommandName=="selectBrand")
            {
Response.Redirect("~/LightListForUsers.aspx?BrandId="+e.CommandArgument);
            }
if(e.CommandName=="selectStyle")
            {
```

```
            Response.Redirect("~/LightListForUsers.aspx?StyleId="+e.CommandArgument);
            }
if(e.CommandName=="selectTopSale"||e.CommandName=="detail")
            {
                StatisticsBll stBll=new StatisticsBll();
if(stBll.AddStoreCount(e.CommandArgument.ToString()))
Response.Redirect("~/DetailInfo.aspx?LightId="+e.CommandArgument);
            }
if(e.CommandName=="selectCategory")
            {
Response.Redirect("~/LightListForUsers.aspx?CategoryId="+e.CommandArgument);
            }
        }
    }
}
```

主页面上部对应的实现代码：Site.Master.cs

```
namespace MyLightShop
{
public partial class SiteMaster : System.Web.UI.MasterPage
    {
protected void Page_Load(object sender, EventArgs e)
        {
string time ="" + DateTime.Now.Year +"." + DateTime.Now.Month +"." + DateTime.Now.Day;
            LabelDate.Text=time;
if(!IsPostBack){ }
        }
protected void LoginLinkButton_Click(object sender, EventArgs e)
            {
Response.Redirect("~/Login.aspx");
            }
        }
}
```

2. 用户信息管理功能表示层及控制层的设计和实现

1) 用户注册部分表示层及控制层的设计与实现

若用户想要在本商店中购买灯饰品，就必须先注册成为本系统的用户，才能进行收藏商品、将商品加入购物车、购买商品等操作，否则只能在图 5.29 所示的主页面上进行商品信息查询及了解商店信息等操作。注册页面如图 5.30 所示，在页面的对应区域填入相关信息，单击"注册"按钮，如果数据无错误，则可以成功注册成为本系统的新用户。注册功能对应的实现代码在 Register.aspx.cs 中，具体内容如下。

图 5.30 网上灯饰店系统注册界面

```
namespace MyLightShop
{
public partial class Register : System.Web.UI.Page
    {
        UsersBll userBal=new UsersBll();
protected void Page_Load(object sender, EventArgs e)
        {
if(!IsPostBack)
            {
LoginIdTextBox.Focus();
                PasswordTextBox.Text="";
                ConfirmTextBox.Text="";
                AdressTextBox.Text="";
                PhoneTextBox.Text="";
                PostCodeTextBox.Text="";
                EmailTextBox.Text="";
                RealNameTextBox.Text="";
                RadioButtonMale.Checked=true;
            }
if(!string.IsNullOrEmpty(Request["Message"]))
            {
                MessageLabel.Text=Request["Message"].ToString();
                MessageLabel.Visible=true;
            }
        }
        protected void btnSave_Click(object sender, EventArgs e)         //注册事件
        {
            User user=new User();
if(this.PasswordTextBox.Text.Trim()==this.ConfirmTextBox.Text.Trim())
```

```csharp
        {
    if(userBal.IsExist(LoginIdTextBox.Text.Trim()).Rows.Count>0)
            {
                MessageLabel.Text="该用户名已存在,请重新输入";
LoginIdTextBox.Focus();
                PasswordTextBox.Text="";
                ConfirmTextBox.Text="";
                AdressTextBox.Text="";
                PhoneTextBox.Text="";
                PostCodeTextBox.Text="";
                EmailTextBox.Text="";
                RealNameTextBox.Text="";
                RadioButtonMale.Checked=true;
            }
    else
            {
                user.loginId=LoginIdTextBox.Text.Trim();
                user.loginPwd=PasswordTextBox.Text.Trim();
                user.email=EmailTextBox.Text.Trim();
                user.userName=RealNameTextBox.Text.Trim();
                user.postalCode=PostCodeTextBox.Text.Trim();
                user.address=AdressTextBox.Text.Trim();
                user.phone=PhoneTextBox.Text.Trim();
if(RadioButtonMale.Checked)
                    user.gender="男";
                else user.gender="女";
                user.lastLoginTime=GetDate();
if(userBal.Resgister(user))
                {
Response.Redirect("~/Login.aspx?MessageInfo="+"恭喜您,用户注册成功,您现在即可登录");
                }
    else
                {
Response.Redirect("~/Register.aspx?Message="+"非常抱歉,您的注册失败!");
                }
            }
        }
    else
        {
            ErrorMessage.Text="两次密码不一致";
            ConfirmTextBox.Text="";
        }
```

```csharp
        protected void btnReset_Click(object sender, EventArgs e)          //重置事件
        {
            LoginIdTextBox.Text="";
            PasswordTextBox.Text="";
            ConfirmTextBox.Text="";
            AdressTextBox.Text="";
            RadioButtonMale.Checked=true;
            PhoneTextBox.Text="";
            PostCodeTextBox.Text="";
            EmailTextBox.Text="";
            RealNameTextBox.Text="";
        }
        public string GetDate()                          //获得当前日期
        {
int intYear=DateTime.Now.Year;
int intMonth=DateTime.Now.Month;
int intDate=DateTime.Now.Day;
string strTime=null;
strTime=intYear.ToString();
if(intMonth<10)
            {
strTime+="-0"+intMonth.ToString();
            }
else
            {
strTime+="-"+intMonth.ToString();
            }
if(intDate<10)
            {
strTime+="-0"+intDate.ToString();
            }
else
            {
strTime+="-"+intDate.ToString();
            }
return strTime;
        }
    }
}
```

2）登录部分

注册的用户可以在登录页面上输入用户名及密码登录到本系统。

实现代码在 Login.aspx.cs 中。具体代码如下：

```csharp
namespace MyLightShop
```

```csharp
public partial class Login : System.Web.UI.Page
    {
        UsersBll uesrBll=new UsersBll();
        WishListBll wishlist=new WishListBll();
        MyShopcartBll myshopcart=new MyShopcartBll();
protected void Page_Load(object sender, EventArgs e)
        {
            RegisterHyperLink.NavigateUrl="Register.aspx";
if(!IsPostBack)
            {
this.NameTextBox.Focus();
                PasswordTextBox.Text="";
            }
        }
protected void btnReset_Click(object sender, EventArgs e)
        {
            NameTextBox.Text="";
            PasswordTextBox.Text="";
        }
protected void btnLogin_Click(object sender, EventArgs e)
        {
Session["UserId"]=null;
Session["UserName"]=null;
Session["LoginId"]=null;
Session["UserRole"]=null;
            DataTable dataTable = uesrBll.UserLogin(NameTextBox.Text.Trim(),
            PasswordTextBox.Text.Trim(), Convert.ToInt16(RadioButtonList1.
            SelectedValue));

if(dataTable.Rows.Count>0)
            {
Session["UserId"]=dataTable.Rows[0][0].ToString();
Session["UserName"]=dataTable.Rows[0][1].ToString();
Session["LoginId"]=dataTable.Rows[0][9].ToString();
Session["UserRole"]=dataTable.Rows[0][7].ToString();
uesrBll.UpdateLastLoginTime(Session["UserId"].ToString(),GetDate());
if(!string.IsNullOrEmpty(Request["LightIdStore"]))
Response.Redirect("~/ManagerDefault.aspx?LightIdStore="+Request["LightIdStore"].
ToString());
else if(!string.IsNullOrEmpty(Request["LightIdBuy"]))
Response.Redirect("~/ManagerDefault.aspx?LightIdBuy=" + Request["LightIdBuy"].
ToString());
else if(!string.IsNullOrEmpty(Request["LightIdComment"]))
```

```csharp
                Response.Redirect("~/DetailInfo.aspx?LightIdComment=" + Request
["LightIdComment"].ToString());
            else if (!string.IsNullOrEmpty(Request["LightIdDetail"]) && !string.
IsNullOrEmpty(Request["Quantity"]))
                Response.Redirect("~/DetailInfo.aspx?LightId=" + Request["LightIdDetail"].
ToString()+"&Quantity="+Request["Quantity"].ToString());
            else Response.Redirect("~/ManagerDefault.aspx");
        }
        else
        {
            Response.Write(CommonClass.MessageBox("您输入的用户名或密码有误，
            请核对后再重新登录!"));
        }
    }

public string GetDate()
    {
        int intYear=DateTime.Now.Year;
        int intMonth=DateTime.Now.Month;
        int intDate=DateTime.Now.Day;
        string strTime=null;
        strTime=intYear.ToString();
        if(intMonth<10)
        {
            strTime+="-0"+intMonth.ToString();
        }
        else
        {
            strTime+="-"+intMonth.ToString();
        }
        if(intDate<10)
        {
            strTime+="-0"+intDate.ToString();
        }
        else
        {
            strTime+="-"+intDate.ToString();
        }
        return strTime;
    }
}
```

3) 用户个人信息管理(修改个人信息)

注册过的用户登录系统后,单击"个人信息管理"目录按钮,即可在如图 5.31 所示的用户信息管理页面上可查看到自己的基本信息。如果个人信息有变动,可以根据需要修改自己的电话、地址、邮编等变动的信息。如果只是想查看个人信息,之后想返回用户主页,单击"返回"按钮或者单击"首页"目录按钮,均可以返回到用户主页,从而进行其他功能操作。

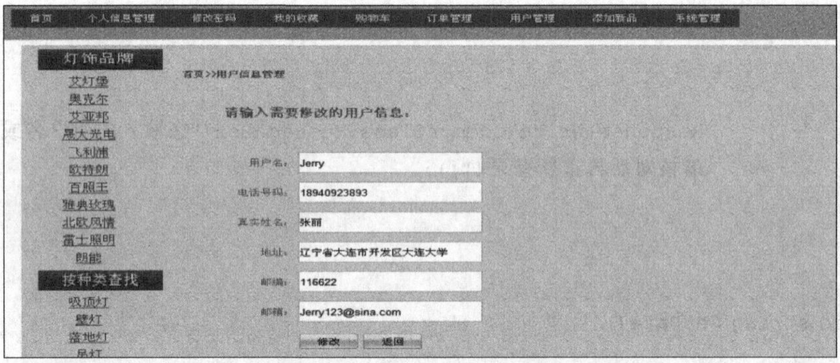

图 5.31 网上灯饰店系统用户信息管理界面

用户个人信息管理实现代码(UserInfo.aspx.cs)如下:

```
namespace MyLightShop
{
public partial class UserInfo : System.Web.UI.Page
    {
        UsersBll userBal=new UsersBll();
protected void Page_Load(object sender, EventArgs e)
        {
if(!IsPostBack)
            {
                DataTable dataTable = userBal. IsExist (Session [ " LoginId "].
                ToString());
                this.NameTextBox.Text=dataTable.Rows[0][9].ToString();
                this.EmailTextBox.Text=dataTable.Rows[0][5].ToString();
                this.PhoneTextBox.Text=dataTable.Rows[0][4].ToString();
                this.PostCodeTextBox.Text=dataTable.Rows[0][3].ToString();
                this.AddressTextBox.Text=dataTable.Rows[0][2].ToString();
                this.RealNameTextBox.Text=dataTable.Rows[0][1].ToString();
            }
        }
protected void btnModify_Click(object sender, EventArgs e)
        {
            User user=new User();
            user.phone=PhoneTextBox.Text.Trim();
```

```
                user.userName=RealNameTextBox.Text.Trim();
                user.loginId=NameTextBox.Text.Trim();
                user.postalCode=PostCodeTextBox.Text.Trim();
                user.email=EmailTextBox.Text.Trim();
                user.address=AddressTextBox.Text.Trim();
                Boolean flag=userBal.Modify(user);
if(flag)
                Response.Write(CommonClass.MessageBox("用户信息修改成功!","
                UserInfo.aspx"));
else
                Response.Write(CommonClass.MessageBox("您输入的信息有误,请确认输
                入的信息格式正确!"));
        }
protected void btnReturn_Click(object sender, EventArgs e)
        {
Response.Redirect("~/ManagerDefault.aspx");
        }
    }
}
```

4) 修改密码

如果需要修改密码,可以单击用户主界面上部的目录栏中的"修改密码",则可进入密码修改界面,如图 5.32 所示。如果旧密码输入正确,而且两次输入的新密码一致,则可以成功修改用户密码。

图 5.32　网上灯饰店系统修改密码界面

修改密码实现部分代码(UpdatePwd.aspx.cs):

```
namespace MyLightShop
{
public partial class UpdatePwd : System.Web.UI.Page
    {
        UsersBll userBal=new UsersBll();
protected void Page_Load(object sender, EventArgs e)
```

```csharp
    {
        if(!IsPostBack)
            this.OldPwdTextBox.Focus();
    }
    protected void btnLogin_Click(object sender, EventArgs e)
    {
        User user=new User();
        DataTable dataTable=userBal.IsExist(Session["LoginId"].ToString());
        user.loginPwd=dataTable.Rows[0][10].ToString();
        user.userId=int.Parse(dataTable.Rows[0][0].ToString());
        if(OldPwdTextBox.Text.Trim()!=user.loginPwd)
        {
            ErrorMessage1.Text="密码输入有误!请重新输入!";
            OldPwdTextBox.Text="";
        }
        if(NewPwdTextBox.Text.Trim()==AffirmPwdTextBox.Text.Trim())
        {
            user.loginPwd=NewPwdTextBox.Text.Trim();
            Boolean flag=userBal.UpdatePwd(user.userId,user.loginPwd);
            if(flag)
            {
                Response.Write(CommonClass.MessageBox("密码修改成!",
                "ManagerDefault.aspx"));
            }
            else
            {
                Response.Write(CommonClass.MessageBox("您输入的密码有误,请确
                认输入的密码正确!"));
            }
        }
        else {
            ErrorMessage2.Text="两次密码不一致";
            NewPwdTextBox.Text="";
            AffirmPwdTextBox.Text="";
        }
    }
    protected void btnReset_Click(object sender, EventArgs e)
    {
        OldPwdTextBox.Text="";
        NewPwdTextBox.Text="";
        AffirmPwdTextBox.Text="";
```

            }
        }
}

5）用户信息删除

如果是以操作员的身份登录，则可以查看到三个月或三个月以上时间未登录系统的用户基本信息，如图 5.33 所示。如果长时间未登录系统的用户数量很多，操作员有权将这些用户信息删除，这样做可以保证数据库中数据的有效性。

图 5.33　操作员的用户管理页面

实现代码：UpdatePwd.aspx.cs

⋮

```
namespace MyLightShop
{
public partial class UpdatePwd : System.Web.UI.Page
    {
        UsersBll userBal=new UsersBll();
protected void Page_Load(object sender, EventArgs e)
        {
if(!IsPostBack)
this.OldPwdTextBox.Focus();
        }

protected void btnLogin_Click(object sender, EventArgs e)
        {
            User user=new User();
            DataTable dataTable=userBal.IsExist(Session["LoginId"].ToString());
            user.loginPwd=dataTable.Rows[0][10].ToString();
            user.userId=int.Parse(dataTable.Rows[0][0].ToString());
```

```
if(OldPwdTextBox.Text.Trim()!=user.loginPwd)
    {
        ErrorMessage1.Text="密码输入有误!请重新输入!";
        OldPwdTextBox.Text="";
    }
if(NewPwdTextBox.Text.Trim()==AffirmPwdTextBox.Text.Trim())
    {
        user.loginPwd=NewPwdTextBox.Text.Trim();
        Boolean flag=userBal.UpdatePwd(user.userId,user.loginPwd);

if(flag)
        {
            Response.Write(CommonClass.MessageBox("密码改成功!","ManagerDefault.aspx"));

        }
    else
        {
            Response.Write(CommonClass.MessageBox("您输入的密码有误,请确认输入的密码正确!"));
        }
    }
else {
        ErrorMessage2.Text="两次密码不一致";
        NewPwdTextBox.Text="";
        AffirmPwdTextBox.Text="";
    }
}
protected void btnReset_Click(object sender, EventArgs e)
    {
        OldPwdTextBox.Text="";
        NewPwdTextBox.Text="";
        AffirmPwdTextBox.Text="";
    }
}
```

3. 商品信息管理功能表示层及控制层设计与实现

商品信息管理功能主要是对系统中的商品信息进行管理,包括对商品信息的查询、按条件搜索符合条件的商品信息、将商品加入收藏夹、将商品添加到购物车、购买商品、评价商品及管理员添加新品的相关功能的管理。

用户登录与否都可以查看商品信息或搜索符合条件的商品信息,但是除此之外的其他操作都需要用户先登录系统才能进行。用户收藏夹和购物车中商品的信息只要是用户不删除,数据库就会一直记录在相应的数据表中,用户登录就可以查看到自己的收藏记录和购物车添加的商品信息。

操作员拥有普通用户可以进行操作的所有与商品信息相关的功能权限,并且添加新品功能是系统管理员才具有的权限,只有管理员才能对该功能进行操作。

1) 按品牌/种类/风格搜索商品信息

用户可以单击主页面左部的导购栏,按商品品牌、商品种类、商品风格搜索不同品牌、种类、风格的商品,如单击"艾亚邦"这个灯饰品牌,会在页面右部显示所有"艾亚邦"品牌的灯饰,如图 5.34 所示。除外,用户还可以在导购栏最下方单击宝贝排行榜中的宝贝,查看商店中最畅销的商品的商品详情。在商品详情中,用户可以查看商品的评价信息以及商品的成交记录信息。

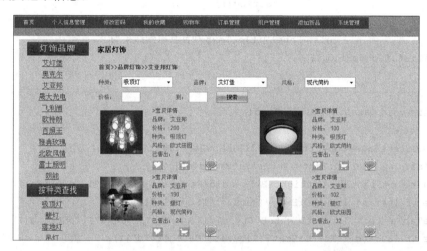

图 5.34　按品牌查询商品信息的结果(艾亚邦品牌灯饰信息)

按品牌/种类/风格查询灯饰的实现代码(LightListForSearch.aspx.cs)如下:

```
namespace MyLightShop
{
public partial class LightListForSearch : System.Web.UI.Page
    {
        DataTable dataTable=null;
        DataTable dataTableForClass=new DataTable();
        BrandBll brandBll=new BrandBll();
        LightBll lightBll=new LightBll();
        StyleBll styleBll=new StyleBll();
        CategoryBll categoryBll=new CategoryBll();
protected static PagedDataSource ps=new PagedDataSource();
protected void Page_Load(object sender, EventArgs e)
        {
if(!IsPostBack)
            {
```

```csharp
if(!string.IsNullOrEmpty(Request["BrandId"]))
            LabelTitle.Text="首页品牌灯饰" + brandBll.GetBrandById
            (Request["BrandId"]).Rows[0][1].ToString()+"灯饰";
else
if(!string.IsNullOrEmpty(Request["StyleId"]))
            LabelTitle.Text="首页灯饰风格" + styleBll.GetStyleById
            (Request["StyleId"]).Rows[0][1].ToString()+"风格灯饰";
else
if(!string.IsNullOrEmpty(Request["CategoryId"]))
            LabelTitle.Text=" 首页灯饰种类 > >" + categoryBll.
            GetCategoryById(Request["CategoryId"]).Rows[0][1].
            ToString();

            else LabelTitle.Text="首页搜索结果";
DataListofGoodsBind();
        }
    }
public void DataListofGoodsBind()
        {
int currentPage=Convert.ToInt32(labNowPage.Text);
if(!string.IsNullOrEmpty(Request["BrandId"]))
dataTable=lightBll.GetLightsByBrandId(Request["BrandId"].ToString());
else if(!string.IsNullOrEmpty(Request["StyleId"]))
dataTable=lightBll.GetLightsByStyleId(Request["StyleId"].ToString());
else if(!string.IsNullOrEmpty(Request["CategoryId"]))
dataTable=lightBll.GetLightByCatagory(Request["CategoryId"].ToString());
else if(!string.IsNullOrEmpty(Request["Sql"]))
dataTable=lightBll.GetLightByInfo(Request["Sql"].ToString());
if(dataTable==null)
            LabelMessage.Visible=true;
else
        {
            ps.DataSource=dataTable.DefaultView;
            ps.AllowPaging=true;
            ps.PageSize=10;
            ps.CurrentPageIndex=currentPage-1;       //取得当前页的页码
            lbkbtnFront.Enabled=true;
            lbkbtnFirst.Enabled=true;
            lbkbtnNext.Enabled=true;
            lbkbtnLast.Enabled=true;
if(currentPage==1)
            {
                lbkbtnFirst.Enabled=false;         //不显示第一页按钮
                lbkbtnFront.Enabled=false;         //不显示上一页按钮
```

```
            }
    if(currentPage==ps.PageCount)
            {
                    lbkbtnNext.Enabled=false;           //不显示下一页按钮
                    lbkbtnLast.Enabled=false;           //不显示最后一页按钮
            }
            this.labCount.Text=Convert.ToString(ps.PageCount);
            DataList2.DataSource=ps;
            DataList2.DataKeyField="LightId";
DataList2.DataBind();
        }
    }
protected void lbkbtnFirst_Click(object sender, EventArgs e)
        {
            this.labNowPage.Text="1";
DataListofGoodsBind();
        }
protected void lbkbtnFront_Click(object sender, EventArgs e)
        {
            this. labNowPage. Text = Convert. ToString (Convert. ToInt32 (this.
            labNowPage.Text)-1);
DataListofGoodsBind();
        }
protected void lbkbtnNext_Click(object sender, EventArgs e)
        {
            this. labNowPage. Text = Convert. ToString (Convert. ToInt32 (this.
            labNowPage.Text)+1);
DataListofGoodsBind();
        }
protected void lbkbtnLast_Click(object sender, EventArgs e)
        {
            this.labNowPage.Text=this.labCount.Text;
DataListofGoodsBind();
        }
protected void ChangeButton_Click(object sender, EventArgs e)
        {
int myPageNum=1;
if(!txtPage.Text.Equals(""))
            {
myPageNum=Convert.ToInt32(txtPage.Text.ToString());
            }
if(myPageNum<=0 || myPageNum>ps.PageCount)
            {
                Response.Write(CommonClass.MessageBox("请输入页数并确定没有超过
```

```
                            总页数"));
            }
        else
            {
                this.labNowPage.Text=Convert.ToString(myPageNum);
DataListofGoodsBind();
            }
        }
    protected void DataList2_ItemCommand(object source, DataListCommandEventArgs e)
        {
if(e.CommandName=="detail")
            {
Response.Redirect("~/DetailInfoDefault.aspx?LightId="+e.CommandArgument);
            }
if(e.CommandName=="buy")
            {
                Response.Write(CommonClass.MessageBox("您还未登录,请先登录!"));
                Response. Redirect ( " ~/Login. aspx? LightIdBuy =" + e.
                CommandArgument);
            }
if(e.CommandName=="store")
            {
                Response.Write(CommonClass.MessageBox("您还未登录,请先登录!"));
                Response. Redirect ( " ~/Login. aspx? LightIdStore =" + e.
                CommandArgument);
            }
if(e.CommandName=="comment")
            {
                Response.Write(CommonClass.MessageBox("您还未登录,请先登录!"));
Response.Redirect("~/Login.aspx?LightIdComment="+e.CommandArgument);
            }
        }
    protected void ButtonSearch_Click(object sender, EventArgs e)
        {
string category=CategoryListBox.SelectedValue;
string style=StyleListBox.SelectedValue;
string brand=BrandListBox.SelectedValue;
string low=TextBoxLow.Text.Trim();
string high=TextBoxHigh.Text.Trim();
string sql="";
if(!low.Equals("")&& !high.Equals(""))
sql="select * from [View_LightInfo] where [CatagoryName]='"+category+"' and
[StyleName]='"+style+"' and [BrandName]='"+brand+"' and [Price] between '"+
int.Parse(low)+"' and '"+int.Parse(high)+"';";
```

```
else sql="select * from [View_LightInfo] where [CatagoryName]='"+category+"'
and [StyleName]='"+style+"' and [BrandName]='"+brand+"';";

Response.Redirect("~/LightListForDefault.aspx?Sql="+sql);
        }
    }
}
```

2）按给定条件查询灯饰

用户可以在主页右部最上方选择商品品牌、种类、风格并输入价格区间，查询满足自己要求的商品信息，如品牌选择雅典玫瑰，种类选择吊灯，风格选择现代简约，价格输入200和600，单击"搜索"按钮，搜索结果在主页右部显示，如图5.35所示。

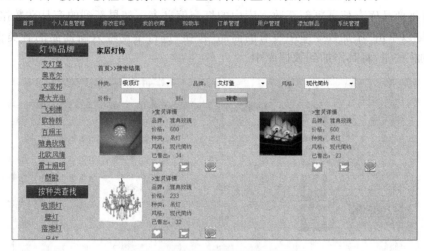

图 5.35　按给定条件搜索结果界面

3）添加商品到收藏夹、购物车

登录系统后，用户在浏览商品信息时，看到自己喜欢的商品但是又不想立刻购买，或者想要再查找一下更满意的商品，则可将该商品添加到自己的收藏夹或购物车中。若想将商品添加到收藏夹中只需要单击商品下方带心的图标即可，而将商品添加到购物车中需要单击商品下方中间左边带小车的图标或者在收藏夹中将收藏的商品"加入购物车"；如果用户想要对某一商品做出自己的评价，还可以单击商品下方右边的评论图标。如果用户已决定要买某件商品，可以在查看商品详情时，单击商品图片右下方的"购买"按钮，或在将该商品添加到购物车之后，单击"购买"按钮则可进入订单管理页面。购物车内容的显示页面如图5.36所示。

4）管理员添加新的商品

在添加新品页面，管理员可以查看到所有库存量低于10件的商品信息（如图5.37所示），这样可以方便管理员对需要补货的商品进行管理，及时补充商品，并修改库存量。操作员可以将新到货物的信息添加到数据库中。管理员可先根据商品编号或者商品的品牌查询商品信息，再在如图5.37所示的"添加数量"那一列中输入添加的数量，单击"添加"

图 5.36 单击购买图标后的界面

按钮即可将商品库存信息更新至数据库中的灯饰信息表。如果新到的商品在数据库中没有记录,可以单击"添加数据库中没有的商品"按钮,进行手动输入新商品的各类信息,再单击"添加"按钮,将其添加到数据库中。

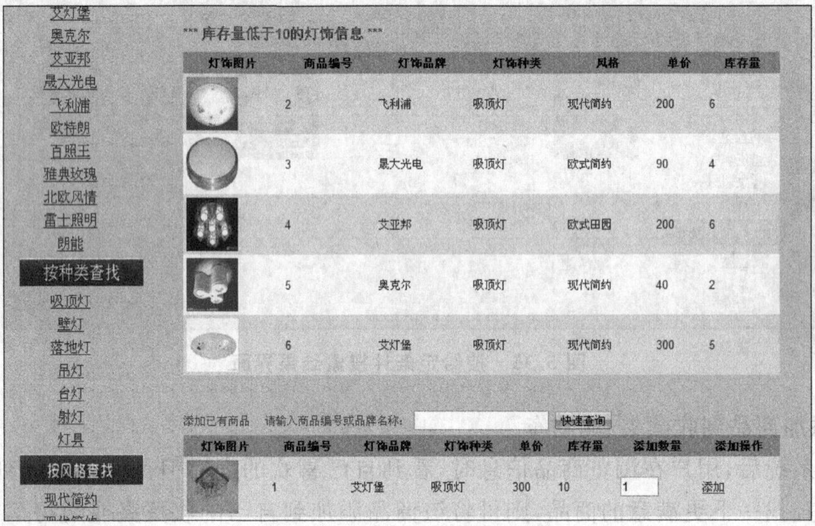

图 5.37 添加新商品界面

4. 订单信息管理功能表示层及控制层设计与实现

订单信息管理功能主要是针对用户的订单信息进行管理,包括对未结算订单进行查询、修改、删除或结算,对已结算未发货订单进行查询,管理员对已结算订单进行查询及发货操作,用户对已发货订单进行查询,若已收到货物,可对订单进行确认收货操作,还可对商品或卖家做出自己的评价。同时用户还可以对自己的历史订单进行查询或清理操作。

1) 查询订单信息

订单信息查询包括对未结算订单、待审核订单、已发货订单及历史订单的查询。

在用户登录后,单击"订单管理"用户可看到自己所有的订单信息。图 5.38 所示的是未结算的订单信息。未结算订单信息左上方显示订单总价。如果用户现在想要处理某一未结算的订单可以单击"去结算"按钮则可进入结算页面,在此操作之前,用户可以修改购买数量。若用户现在不想要该订单了,可进行删除操作。

图 5.38　订单信息页面

2) 结算操作

在显示页面中,用户在单击"订单管理"显示信息(如图 5.38 所示),在图 5.38 中针对未结算某一订单单击"去结算"按钮就可以进入结算页面,如图 5.39 所示。在结算页面填入相关信息单击"提交"按钮,就可进行结算。如果真正地将本系统应用到商店的运营模式中,需要修改"提交"按钮的相关代码,让其链接到第三方支付平台就可以实现真正的支付。

图 5.39　结算页面

3) 发货和确认收货操作

操作员单击"订单管理"目录,就可以进入订单管理页面,操作员可以查看到所有等待配货的订单信息,如图 5.40 所示,单击"发货"按钮即可对该订单发货。发完货的订单会在用户查看订单信息时,显示在等待确认的订单信息表中。如果用户已收到商家发出的货物,登录系统后,在订单管理页面中的等待确认的订单信息表中对收到的商品进行"确认收货"操作,这样就可以将订单的状态置为已接收。在确认完收货之后,用户可以在确认收货页面右下部出现的评价框中填写自己对商品和商家的评价。

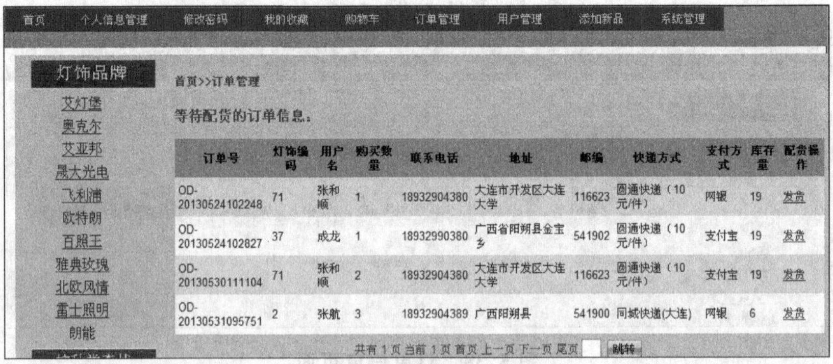

图 5.40　管理员查看待配货订单信息页面

5. 系统管理功能

系统管理功能是操作员才有权进行操作的功能,该功能实现的是将系统中数据库备份到指定存储位置,目的是为了在数据库出故障后,可以利用数据库备份以及数据库的日志文件使数据库恢复到故障前某一时刻的一致性状态。从而保证数据的一致性以及系统的稳定性和安全性。

在系统管理功能中,操作员还可以查看本月的成交记录,如果有需要还可以将这些记录信息导出到 Excel 表格中,操作员只需填写好保存路径,然后单击"导出表格"按钮,即可将记录信息导出到指定路径中。系统管理功能页面如图 5.41 所示。

图 5.41　系统管理功能页面

第 6 章 家具网站的研究与实现

本章介绍的家具网站不是一个实际的建设项目,系统目标和功能是研究人员参照一般购物网站、结合教学过程中的实际情况而设定的。该网站的功能虽然不是很全面,但购物网站建设是按照软件工程的思想、基本按照生命周期法的开发过程进行的,同时结合了面向对象的开发方法(因采用的Java开发环境),过程描述清晰,在开发购物网站过程中遇到的核心问题、主要问题,都能在本章找到解决方法。

本章介绍的家具网站建设采用MVC框架,软件开发环境如下:
(1) 操作系统　Windows XP。
(2) 开发环境　MyEclipse 8.6。
(3) 数据库　MySQL。

6.1 系统规划

在系统规划阶段主要完成如下工作:
(1) 系统需求分析和目标设定;
(2) 对设定的目标进行可行性分析。

6.1.1 系统需求分析和目标设定

根据现有购物网站特别是家具网站的研究,梳理出对系统的需求:
(1) 建立一个相对完善的网上购物系统;
(2) 在网站中相关信息表达准确、显示方式恰当、布局合理;
(3) 记载的信息准确,方便使用;保证信息的安全;
(4) 功能全面;
(5) 方便一般用户完成相关操作,各种操作合乎用户的一般习惯;
(6) 方便管理人员完成相关操作,各种操作合乎管理员的操作习惯;
(7) 各种操作流程合理;
(8) 为相关的人员提供进行决策的有效数据;
(9) 提高销售量。

对上述需求进行进一步分析、整理,提出具体的系统目标为:
(1) 梳理出先进、规范的业务流程;
(2) 快速、准确的信息收集,能真正实现决策支持;
(3) 通过对网站的管理,能逐步提高购物效率和销售量;
(4) 加强对整个购物过程的监控力度,保证系统和数据安全;
(5) 实现数据的集中化、数字化处理。

为了到达上述目标,系统要实现的主要功能如下。

(1) 用户管理:该模块使网站管理员可以通过浏览器在线对用户信息进行管理,以便更好地管理用户;用户可以通过浏览器在线进行注册,成为网站会员,享有相关的权益、履行相关义务。

(2) 建立完善的不同用户身份的权限管理:不同权限拥有不同功能和操作权限,以便更好地管理用户、保障网站能更好地运行。

(3) 商品管理:该模块使网站管理员可以通过浏览器在线添加、修改、删除商品信息,商品的数据可以即时更新,保证用户浏览到最新的商品信息;用户能够对商品进行查询、搜索或购买自己喜欢的商品,对商品进行更新或删除商品信息。

(4) 销售统计:该模块使网站管理员可以通过浏览器统计指定时间段的、各类商品的销售数据,以便更好地进行相关管理或采取相关措施。

(5) 订单管理:该模块使网站管理员可以通过浏览器管理订单,如查看订单状态、执行订单、查询订单等。

(6) 公告管理:该模块使网站管理员可以通过浏览器管理公告,如发布公告、删除公告等,使客户可以及时了解网站和公司的各类情况和信息。

(7) 购物车:用户选定要购买的商品后,单击我的购物车,把要购买的商品添加到购物车中,结账前可以再次确认购买的商品,再单击结账则进入结账功能,从而减少了所卖商品不如意再退货的情况。

6.1.2 系统可行性分析

对上述系统目标和要实现的主要功能在经济、技术和社会三方面进行可行性分析。

1. 经济可行性分析

进行经济可行性主要从投入和收益两方面进行。

开发本网站投入的部分较少:一台比较高档的个人计算机即可,开发系统一个人用半年时间就可完成网站建设软件部分;在运行阶段需要管理人员(专职)人数视情况而定,维护人员(可兼职)一人即可。

网站建设完成之后,可以扩大企业的销售范围而不用增加销售人员,可以全年无休日进行销售。节省了成本和花销(包括人工、店面等的成本和花销),收益很大。

2. 技术可行性分析

硬件方面:需要比较高档的个人计算机在市面上可以很容易买到。

软件方面:建设网站使用的数据库是 MySQL,开发平台是 MyEclipse。

人员方面：需要的开发人员具备一定的开发网站的经验、会使用 MySQL、MyEclipse 以及能用 Java 语言编程。

3．社会可行性分析

开发网站，利用网站进行企业宣传和产品销售，是电子商务的一部分，是国家大力提倡的，因此建设网站在社会方面是可行的。

6.2 系统分析

在系统分析阶段主要完成如下工作。

1．业务流程分析与描述

利用业务流程图描述如下主要业务：

（1）销售业务；

（2）订单管理业务；

（3）用户管理业务；

（4）公告管理业务；

（5）商品管理业务。

2．数据流程分析与描述

利用数据流程图描述如下数据流程：

（1）系统顶层数据流程；

（2）系统第一层数据流程；

（3）系统第二层中的用户相关功能数据流程；

（4）系统第二层中的操作员相关功能数据流程。

3．数据分析与描述

利用表格对各数据项、外部实体、各数据流、各数据存储以及各加工处理进行了描述，即建立数据流图中的各种对象的数据字典。

6.2.1 业务流程分析与描述

为了达到系统设定的目标，实现系统要求的功能，经过分析梳理出完成这些功能应具备的相关的业务，主要包括销售业务、订单管理业务、用户管理业务、公告管理业务以及商品管理业务。

1．销售业务

销售简单来说就是卖出家具。在这项业务中涉及普通用户和操作员。普通用户通过浏览家具对想购买的家具进行选择，把选择的家具放入购物车，在购物车中的家具可以进行二次选择，确定购买之后生成订单，再由操作员确认订单、执行订单；操作员对订单进行相关统计即进行销售统计。这项业务涉及的数据存储有家具信息、订单存储，涉及的表有购物车表和订单表。详细的业务流程过程如图 6.1 所示。

2．订单管理业务

在这项业务中涉及普通用户和操作员。普通用户浏览自己的订单;操作员针对用户填写的订单进行确认订单、执行订单和做相关的销售统计。这项业务涉及的数据存储为订单存储。详细的业务流程过程如图 6.2 所示。

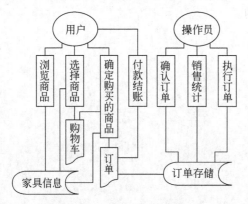

图 6.1　销售业务流程图

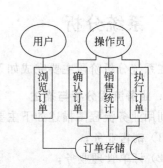

图 6.2　订单管理业务流程图

3．用户管理业务

在这项业务中涉及普通用户和操作员。普通用户可以进行登录、注册和个人信息修改;操作员审查用户填写的信息(可由系统完成)、查看注册的用户和删除注册的用户。这项业务涉及的数据存储为用户档案,涉及的表为用户信息表。详细的业务流程过程如图 6.3 所示。

4．公告管理业务

在这项业务中涉及普通用户和操作员。普通用户可以浏览公告;操作员可以浏览公告,同时可以完成添加公告、删除公告的工作。这项业务涉及的数据存储为公告档案。详细的业务流程过程如图 6.4 所示。

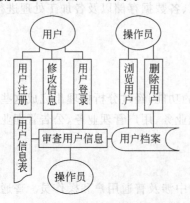

图 6.3　用户管理业务流程图

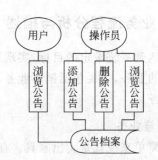

图 6.4　公告管理业务流程图

5．商品管理业务

在这项业务中涉及普通用户和操作员。普通用户可以浏览商品;操作员可以浏览商品,同时可以完成添加公告、修改商品和删除商品的工作。这项业务涉及的数据存储为商品档案,涉及的表为商品信息表。详细的业务流程过程如图6.5所示。

6.2.2 数据流程分析与描述

数据流程的分析就是对系统需求和业务流程进行进一步分析,对其进行进一步抽象,把其中单据、物质、资金等移动或转移或传递抽象为信息流,处理的业务和各业务处理环节抽象为加工处理,涉及到的部门和人员抽象为外部实体,各种档案等抽象为数据存储,把分析和抽象的结果利用数据流图进行描述。

经过对业务流程进行进一步分析和抽象,得到家具网站的顶层数据流程,对应的数据流图如图6.6所示。从图6.6中可以看到家具网站的总处理用"家具网站"名称描述,涉及四项数据存储,把外部实体统称为用户,五种数据流。

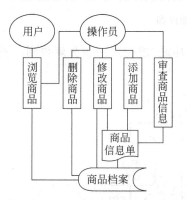
图6.5 商品管理业务流程图

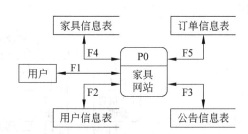
图6.6 家具网站顶层数据流图

参照业务流程和系统功能,把顶层数据流图中的功能进行进一步分解,在分解处理的同时把外部实体、数据流和数据存储也进行进一步分解,得到家具网站的第一层数据流程,对应的数据流图如图6.7所示。从图6.7中可以看到,把"家具网站"处理功能分解成"用户登录"、"浏览家具"、"查询家具"、"浏览公告"、"普通用户相关功能"、"操作员相关功能"共六种处理功能。在分解功能的同时把外部实体也进行了分解,分解成"用户"、"普通用户"和"操作员"。数据流也都进行了进一步分解,F1分解成六个数据流,分别为F1.1、F1.2、F1.3、F1.4、F1.5和F1.6;F2分解成三个数据流,分别为F2.1、F2.2和F2.3;F3分解成两个数据流,分别为F3.1和F3.2;F4分解成四个数据流,分别为F4.1、F4.2、F4.3和F4.4;F5分解成两个数据流,分别为F5.1和F5.2;增加了F6数据流;数据存储增加了购物车表。

参照业务流程和系统功能,把第一层数据流图中的功能进行进一步分解,在分解处理的同时把外部实体、数据流和数据存储也进行进一步分解,得到家具网站的第二层数据流程,由于第二层数据流程比较详细,如果用一张数据流图描述,这张图比较大、看起来也比较乱,为此按第一层数据流图处理加工为单位,描述第二层数据流程,即第一层数据流图中的每一个处理对应一张第二层数据流图。在此给出了第一层数据流图中的"普通用户相关功能"和"操作员相关功能"的第二层数据流图,图6.8为对"普通用户相关功能"进行

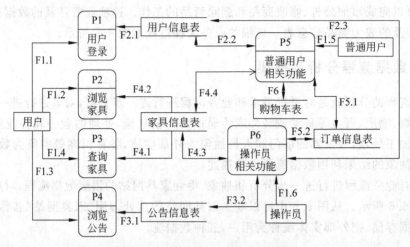

图 6.7 家具网站第一层数据流图

进一步分解得到的数据流图,图 6.9 为对"操作员相关功能"进行进一步分解得到的数据流图。

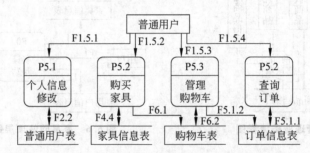

图 6.8 第二层数据流图(普通用户相关功能)

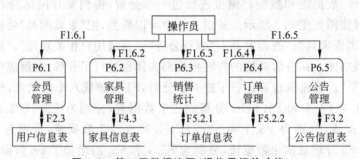

图 6.9 第二层数据流图(操作员相关功能)

从图 6.8 中可以看到,把"普通用户相关功能"的处理功能分解成"个人信息修改"、"购买家具"、"管理购物车"和"查询订单"共四种处理功能。在分解功能的同时把数据流也都进行了进一步分解,F1.5 分解成四个数据流,分别为 F1.5.1、F1.5.2、F1.5.3 和 F1.5.4;F5.1 分解成两个数据流,分别为 F5.1.1 和 F5.1.2;F6 分解成两个数据流,分别为 F6.1 和 F6.2。

从图 6.9 中可以看到,把"操作员相关功能"的处理功能分解成"会员管理"、"家具管理"、"销售统计"、"订单管理"和"公告管理"共五种处理功能。在分解功能的同时把数据流也都进行了进一步分解，F1.6 分解成五个数据流，分别为 F1.6.1、F1.6.2、F1.6.3、F1.6.4 和 F1.6.5；F5.2 分解成两个数据流，分别为 F5.2.1 和 F5.2.2。

6.2.3 数据分析与描述

数据流程进行分析之后，就要对数据流图中的相中元素进行分析和描述，描述的工具为数据字典。数据字典是对数据流图中的数据流、处理逻辑、外部实体、数据存储和数据项等方面进行具体的定义。数据流程图配以数据字典。数据流图和数据字典就是对系统的逻辑模型进行完整的描述。

1. 数据项

数据项是数据流和数据存储中不可再分的数据单位，对数据项的描述一般包括数据项名、数据项含义说明、别名、数据类型、长度、取值范围、取值含义、与其他数据项的逻辑关系以及数据项之间的联系。最后两项描述定义了数据的完整性约束条件，常作为数据校验的依据。

对上述的数据流图中的数据项的描述具体内容，如表 6.1 所示。

表 6.1　家具网站数据项定义表

序号	编号	名 称	别 名	类 型	长度/字节
1	I01	用户编号	cid	int	11
2	I02	用户名	cname	varchar	20
3	I03	用户密码	cpassword	varchar	6
4	I04	电话	ctel	varchar	11
5	I05	金额	cmoney	int	4
6	I06	角色	ostatus	varchar	2
7	I07	操作员编号	oid	int	11
8	I08	操作员密码	opassword	varchar	20
9	I09	姓名	oname	varchar	20
10	I10	角色	ostatus	int	2
11	I11	家具编号	fid	int	11
12	I12	家具名	fname	varchar	20
13	I13	概要	fsummary	varchar	50
14	I14	数量	fnum	int	11
15	I15	单价	fmoney	int	4
16	I16	图片	fpic	varchar	40
17	I17	购物车编号	ID	int	11

续表

序号	编号	名称	别名	类型	长度/字节
18	I18	订单编号	orderID	int	11
19	I19	单价	price	int	11
20	I20	数量	number	int	11
21	I21	品种数	bnum	int	11
22	I22	真实姓名	ruename	varchar	20
23	I23	联系地址	address	varchar	100
24	I24	是否执行	enforce	int	4
25	I25	说明	bz	varchar	200
26	I26	公告编号	ID	int	4
27	I27	标题	title	varchar	100
28	I28	内容	content	mediumtext	不限

2．数据结构

数据项按一定方式组成数据结构，数据流和数据存储中包含一项或几项数据结构，对数据结构的描述一般包括数据结构名称和组成数据结构的数据项。

对数据流图中的数据流、数据存储进行分析，结合数据项的定义，最后定义的数据结构如表6.2所示。

表6.2　家具网站数据结构定义表

序号	编号	名称	组成
1	S01	用户登录信息	用户名＋密码＋角色
2	S02	基本用户信息	用户名＋密码＋电话号码
3	S03	家具信息	家具名＋概要＋数量＋单价＋图片
4	S04	购物车信息	购物车编号＋订单编号＋家具编号＋单价＋数量
5	S05	订单信息	订单编号＋品种数＋用户名＋真实姓名＋联系地址＋是否执行＋备注
6	S06	销售统计信息	品种数＋金额＋总金额
7	S07	公告信息	标题＋内容
8	S08	普通用户信息	基本用户信息＋角色＋金额
9	S09	操作员信息	用户名＋密码＋角色

续表

序号	编号	名称	组成
10	S10	浏览操作信息	家具名＋概要＋数量＋单价＋图片
11	S11	购买操作信息	订单编号＋品种数＋用户名＋真实姓名＋联系地址＋是否执行＋备注
12	S12	加入购物车操作信息	购物车编号＋订单编号＋家具编号＋单价＋数量
13	S13	查询订单相关信息	订单编号＋品种数＋用户名＋真实姓名＋联系地址＋是否执行

3．数据流

数据流图中的数据流是动态数据，有来源和去向，在加工处理中被处理，也可以保存在数据存储中，还可以提供给外部实体；数据流是由一项或几项数据结构组成，或数据项组成。对数据流的描述一般包括数据流名称、来源、去向和组成。对数据流图中的数据流进行分析和梳理，结合数据项和数据结构的定义，最后定义的数据流如表6.3所示。

表6.3 家具网站数据流定义表

序号	数据流编号	名称	来源	去向	组成
1	F1		用户	家具网站	
2	F1.1	用户登录信息	用户	用户登录	S01
3	F1.2	浏览家具操作信息	用户	浏览家具	S10
4	F1.3	浏览公告信息操作信息	用户	浏览公告	I30
5	F1.4	家具名	用户	查询家具	I12
6	F1.5		普通用户	普通用户相关功能	
7	F1.5.1	个人信息修改的相关信息	普通用户	个人信息修改	S02
8	F1.5.2	购买家具的相关信息	普通用户	购买家具	S11
9	F1.5.3	购物车管理的相关信息	普通用户	购物车管理	S12
10	F1.5.4	查询订单操作信息	普通用户	订单查询	S13
11	F1.6		操作员	操作员相关功能	S01＋S02＋S03＋S05＋S07
12	F1.6.1	对用户进行管理的信息	操作员	会员管理	S01＋S02
13	F1.6.2	对家具进行管理的信息	操作员	家具管理	S03
14	F1.6.3	时间	操作员	销售统计	S06

续表

序号	数据流编号	名 称	来 源	去 向	组 成
15	F1.6.4	对订单进行管理的信息	操作员	订单管理	S05
16	F1.6.5	对公告进行管理的信息	操作员	公告管理	S07
17	F2		家具网站	用户信息表	S01＋S02＋S09
18	F2.1	用户登录信息	用户登录	用户信息表	S01
19	F2.2	与普通用户功能有关的用户相关信息	普通用户相关功能	用户信息表	S02
20	F2.3	与操作员功能有关的用户相关信息	操作员相关功能	用户信息表	S09
21	F3	公告信息	家具网站	公告信息表	S07
22	F3.1	公告信息	公告信息表	浏览公告	S07
23	F3.2	公告信息	操作员相关功能	公告信息表	S07
24	F4		家具网站	家具信息表	S03
25	F4.1	符合条件的家具信息	查询家具	家具信息表	S03
26	F4.2	家具信息	浏览家具	家具信息表	S03
27	F4.3	家具信息	操作员相关相关功能	家具信息表	S03
28	F4.4	家具信息	普通用户相关功能	家具信息表	S03
29	F5		家具网站	订单信息表	S05
30	F5.1	确定的购物信息＋订单信息	普通用户相关功能（订单信息表）	订单信息表（普通用户相关功能）	S05
31	F5.1.1	确定的购物信息	订单管理（订单信息表）	订单信息表（订单管理）	S05
32	F5.1.2	订单信息	购物车管理	订单信息表	S05
33	F5.2		操作员相关功能（订单信息表）	订单信息表（操作员相关功能）	S05
34	F5.2.1	订单信息	订单管理（订单信息表）	订单信息表（订单管理）	S05
35	F5.2.2	订单信息	销售统计（订单信息表）	订单信息表（销售管理）	S05
36	F6		普通用户相关功能（购物车表）	购物车表（普通用户相关功能）	S04
37	F6.1	购物相关信息	购买家具	购物车表	S04
38	F6.2	购物车相关信息	购物车表（购物车管理）	购物车管理（购物车表）	S04

4．数据存储

数据流图中的数据存储是静态数据，用来临时或长期保存数据，由一项或几项数据结构组成。对数据存储的描述一般包括名称、描述和组成。对数据流图中的数据存储进行分析和梳理，结合数据结构的定义，最后定义的数据存储如表6.4所示。

表6.4　家具网站数据存储定义表

序号	编号	名 称		描 述	组成
1	D01	用户信息表	普通用户表	用来描述普通用户的基本信息	S08
2	D02		操作员表	用来描述操作员的基本信息	S09
3	D03	家具信息表		用来描述家具的基本信息	S03
4	D04	公告信息表		用来描述公告的基本信息	S07
5	D05	购物车表		用来描述用户选定的家具信息	S04
6	D06	订单表		用来描述用户购买的家具信息	S05

5．外部实体

数据流图中的外部实体的描述一般包括名称和描述。对数据流图中的加工处理的描述如表6.5所示。

表6.5　家具网站外部实体定义表

序号	编号	实体名称	描 述
1	W01	用户	包括普通用户和操作员，以及之外打开页面的人员
2	W02	普通用户	在网站上进行注册的用户
3	W03	操作员	网站的管理人员

6．加工处理

数据流图中的加工处理的描述一般包括名称和功能描述。对数据流图中的加工处理描述如表6.6所示。

表6.6　家具网站加工处理定义表

序号	编号	名 称	功能描述
1	P0	家具店系统	网站具备的所有功能，需进一步分解
2	P1	用户登录	进行用户登录和注册
3	P2	浏览家具	显示所有家具
4	P3	查询家具	显示按指定家具名称的家具
5	P4	浏览公告	显示公告标题和显示指定公告的具体内容
6	P5	普通用户功能	以普通用户身份登录系统后拥有的功能

续表

序号	编号	名称	功能描述
7	P5.1	个人信息修改	修改相关的个人信息
8	P5.2	购买家居	把选定购买的家具放入购物车中
9	P5.3	购物车管理	对选定购买的家具进行管理：移去、清空、结账(生成订单)、继续购买等
10	P5.4	订单查询	对订单进行查询
11	P6	操作员用户功能	以操作员身份登录系统后拥有的功能
12	P6.1	会员管理	对会员(已经注册的用户和操作员)进行管理
13	P6.2	家具管理	对家具进行管理，包括修改和删除已存在家具，添加新家具等
14	P6.3	销售统计	对指定时间段的销售的家具进行统计
15	P6.4	订单管理	对普通用户的订单进行管理
16	P6.5	公告管理	对公告进行管理，包括删除已存在公告，添加新公告等

6.3 系统设计

在系统设计阶段主要完成如下工作。

1. 系统功能结构设计

利用功能结构图描述如下主要功能：

(1) 系统总体功能；

(2) 登录注册功能；

(3) 普通用户功能；

(4) 操作员功能。

2. 系统功能详细设计

利用算法描述工具描述如下功能的实现算法：

(1) 登录和注册功能；

(2) 查询家具；

(3) 修改个人信息；

(4) 购买家具；

(5) 购物车；

(6) 管理家具；

(7) 管理用户；

(8) 管理公告；

(9) 管理订单；

（10）销售统计。

3. 数据库设计

在这里数据库设计主要完成概念设计和逻辑设计。

在概念设计中，完成了各实体及实体与实体之间联系的设计，并利用 E-R 图进行描述。描述的实体包括普通用户、操作员、订单、购物车、公告；描述的实体与实体间的联系的有普通用户、订单和购物车三种实体之间的联系、订单和购物车两种实体之间的联系。

在逻辑设计中，依据概念设计的各实体以及实体间的联系，在相关理论的指导下设计出普通用户、操作员、订单、购物车、公告的逻辑结构。

6.3.1 系统设计思想简介

本网站采用的 MVC 框架，总的设计思想是：用户通过前台（用户界面即网页）提供相关信息，经过控制层进行处理，从后台（数据库）提取相关数据或保存相关数据，把处理结果通过前台返给用户。网站系统总体框架如图 6.10 所示。

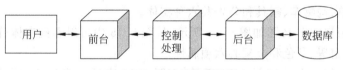

图 6.10 系统总体框架图

6.3.2 系统功能结构设计

对数据流图进行进一步分析，按照系统设计的原则，特别是可由数据流图导出功能结构图，从而设计出系统的总体功能结构。系统总体功能结构可通过系统总体功能结构图表达出来，家具网站的系统总体功能结构图如图 6.11 所示。

依据数据流图对登录注册功能进行分析和梳理，结果描述如图 6.12 所示。

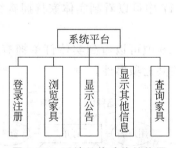 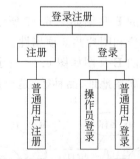

图 6.11 系统总体功能结构图　　图 6.12 登录注册功能结构图

依据数据流图对操作员的功能进行分析和梳理，结果描述如图 6.13 所示。
依据数据流图对普通用户的功能进行分析和梳理，结果描述如图 6.14 所示。

图 6.13 操作人员功能结构图

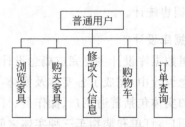

图 6.14 普通用户功能结构图

6.3.3 数据库设计

在这部分中描述的数据库设计分为两部分——概念设计部分和逻辑设计部分。

1. 概念设计

概念设计主要是建立系统的实体——联系模型。

经过系统分析阶段对数据流和数据存储的分析,尤其是数据字典,抽象出了普通用户、操作员、订单、购物车、家具和公告共六种实体。

实体普通用户的 E-R 图如图 6.15 所示,从图 6.15 中可以看到实体普通用户拥有编号、姓名、电话、密码、角色和金额共六项属性。

实体操作员的 E-R 图如图 6.16 所示,从图 6.16 中可以看到实体操作员拥有编号、姓名、密码和角色共四项属性。

图 6.15 普通用户 E-R 图　　　　　图 6.16 操作员 E-R 图

实体家具的 E-R 图如图 6.17 所示,从图 6.17 中可以看到实体家具拥有编号、名称、概要、数量、单价和图片共 6 项属性。

实体订单的 E-R 图如图 6.18 所示,从图 6.18 中可以看到实体订单拥有编号、用户名、真实姓名、联系地址、品种数、是否执行和备注共七项属性。

图 6.17 家具 E-R 图　　　　　图 6.18 订单 E-R 图

实体购物车的 E-R 图如图 6.19 所示，从图 6.19 中可以看到实体购物车拥有编号、单价、数量、订单编号、家具编号和是否执行共六项属性。

实体公告的 E-R 图如图 6.20 所示，从图 6.20 中可以看到实体公告拥有编号、标题和内容共三项属性。

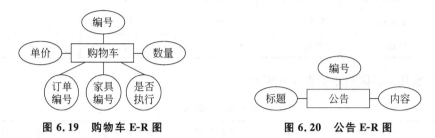

图 6.19　购物车 E-R 图　　　　　图 6.20　公告 E-R 图

在这些实体中，普通用户、家具、购物车之间存在联系，购物车和订单之间存在联系，这些实体之间的联系在图 6.21 中描述。

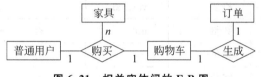

图 6.21　相关实体间的 E-R 图

2．逻辑设计

通过概念设计获得了实体、实体拥有的属性，以及实体之间的联系。在逻辑设计把概念设计的概念模型转换为特定数据模型下的逻辑模型。建设网站采用的关系型数据库，因而把概念模型中实体、实体联系转换为关系模型下的关系。

依据概念模型转化为逻辑模型的原则和方法，结合规范化理论，设计出了普通用户、操作员、家具、订单、购物车和公告共六种关系。

关系普通用户的具体描述如表 6.7 所示。

表 6.7　普通用户表

序号	字段名称	名　称	数据类型	是否可为空	说　明
1	cid	用户编号	int	否	唯一值　主键
2	cname	用户名	varchar	否	
3	cpassword	用户密码	varchar	否	
4	ctel	电话	varchar		
5	cmoney	金额	varchar		
6	ostatus	角色	varchar	否	

关系操作员的具体描述如表 6.8 所示。

表 6.8 操作员表

序号	字段名称	名称	数据类型	是否可为空	说明
1	oid	操作员编号	int	否	唯一值 主键
2	opassword	操作员密码	varchar	否	
3	oname	姓名	varchar	否	
4	ostatus	角色	int	否	

关系家具的具体描述如表 6.9 所示。

表 6.9 家具表

序号	字段名称	名称	数据类型	是否可为空	说明
1	fid	家具编号	int	否	唯一值 主键
2	fname	家具名	varchar	否	
3	fsummary	概要	varchar	否	
4	fnum	数量	int	否	
5	fmoney	单价	int	否	
6	fpic	图片	varchar	否	

关系购物车的具体描述如表 6.10 所示。购物车表,临时存储用户在访问网站时添加至购物车中的商品信息。

表 6.10 购物车表

序号	字段名称	名称	数据类型	是否可为空	说明
1	ID	购物车编号	int	否	唯一值 主键
2	orderID	订单编号	int	否	与订单表的 orderID 关联
3	fid	家具编号	int	否	与商品表 fid 关联
4	price	单价	int	否	
5	number	数量	int	否	

关系订单的具体描述如表 6.11 所示。订单表,将用户确定购买的商品的相关信息保存以便于发货和操作员对订单的管理。

表 6.11 订单表

序号	字段名称	名称	数据类型	是否可为空	说明
1	OrderID	订单编号	int	否	唯一值 主键
2	bnum	品种数	int	否	
3	cname	用户名	varchar	否	

续表

序号	字段名称	名称	数据类型	是否可为空	说明
4	truename	真实姓名	varchar	否	
5	address	联系地址	varchar	否	
6	enforce	是否执行	int	否	是否执行
7	bz	备注	varchar		

关系公告的具体描述如表 6.12 所示。

表 6.12 公告表

序号	字段名称	名称	数据类型	是否可为空	说明
1	ID	编号	int	否	唯一值　主键
2	title	标题	varchar	否	
3	content	内容	mediumtext		

对数数据库设计还有一部分——物理设计，就是把设计出关系转换为具体数据库环境下的表以及表的结构。这些内容在系统实现部分介绍。

6.3.4 系统功能详细设计

系统功能详细设计就是完成功能结构图中的每项功能的具体实现方法——算法，并完成描述。

用户登录的实现算法的描述如图 6.22 所示。主要对用户提供的用户名、密码和角色进行验证。需要访问普通用户表和操作员表。在这里限定输入次数，读者自己可以添加上。

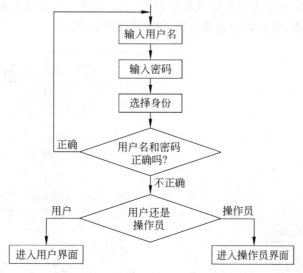

图 6.22 登录功能的程序流程图

用户注册的实现算法的描述如图 6.23 所示。主要对用户提供的用户名进行检查，查看是否重名，不允许重名；检查两次输入的密码是否一致。需要访问普通用户表。

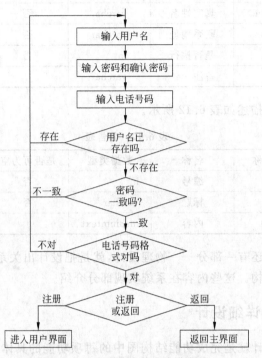

图 6.23　注册功能的程序流程图

查询家具的功能实现比较简单，实现算法如图 6.24 所示，涉及家居信息表。这里只实现了按家具名进行查询，读者可以自己实现按其他项查询家具。

普通用户中的修改个人信息实现算法如图 6.25 所示。主要检查密码是否一致、电话号码的格式是否正确，确认是否修改信息，涉及普通用户表。

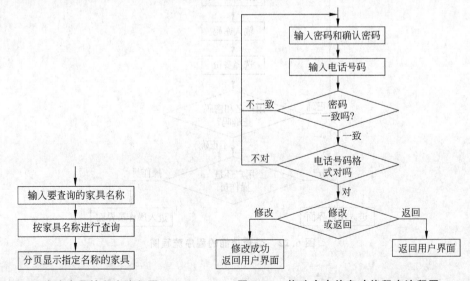

图 6.24　查询家具的程序流程图　　　　**图 6.25　修改个人信息功能程序流程图**

普通用户中的购买家具实现算法如图 6.26 所示。主要是把用户选中的家具添加到用户购物车中。涉及家具信息表和购物车表。

普通用户中的购物车管理实现算法如图 6.27 所示。该实现算法比较复杂,对购物车管理包括清空购物车、从购物车中移去家具、继续购物添加以及结账(即下订单)。涉及家具信息表、订单、普通用户信息表和购物车表。

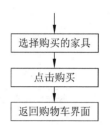

图 6.26 购买家具功能程序流程图

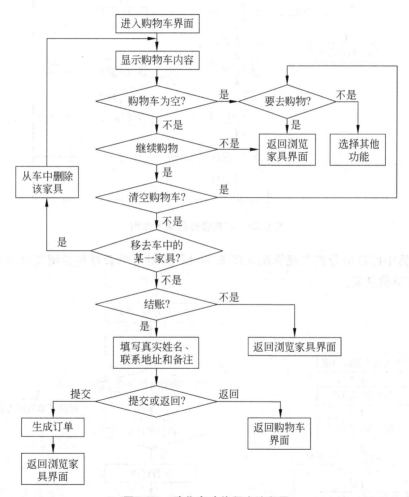

图 6.27 购物车功能程序流程图

操作员中的家具管理实现算法如图 6.28 所示。该实现算法比较复杂,对家具管理包括删除已有家具、添加新家具、修改已有家具。

操作员中的用户管理实现算法如图 6.29 所示,对用户管理主要是删除指定用户,涉及普通用户信息表。

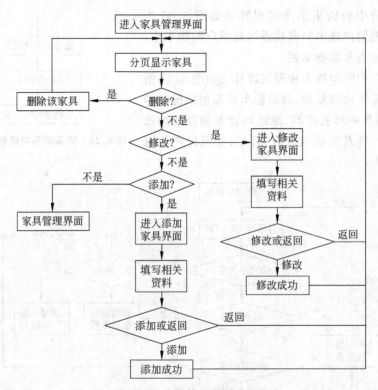

图 6.28 家具管理程序流程图

操作员中的订单管理实现算法如图 6.30 所示,对订单管理包括浏览订单和执行订单,涉及订单信息表。

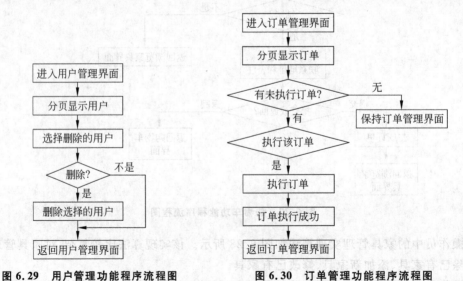

图 6.29 用户管理功能程序流程图　　图 6.30 订单管理功能程序流程图

操作员中的公告管理实现算法如图 6.31 所示,对公告管理包括添加公告和删除公告,涉及公告信息表。

操作员中的销售统计实现算法如图 6.32 所示，主要是对指定时间段中的订单进行数量、金额统计，涉及订单信息表。

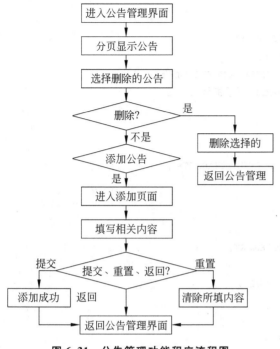

图 6.31 公告管理功能程序流程图

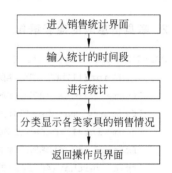

图 6.32 销售统计功能程序流程图

6.4 系统实现

在系统实现与测试阶段主要完成如下工作。

1. 数据库的建立与连接

在 MySQL 中建立了数据库 dalongweb，在该数据库中建立了相关表。建立的表有：
（1）普通用户表；
（2）操作员表；
（3）订单表；
（4）公告表；
（5）购物车表；
（6）家具表。

2. 各模块的实现与测试

编写出各模块的实现代码，并进行测试。

6.4.1 数据库的建立与连接

在 MySQL 中建立了数据库 dalongweb，并建立了与关系对应的六个表，建表的详细语句如下。

1. 普通用户表的建表语句

```
CREATE TABLE `consumer`(
    `cid` int(11)NOT NULL AUTO_INCREMENT,
    `cname` varchar(20)DEFAULT NULL,
    `cpassword` varchar(6)DEFAULT NULL COMMENT '密码',
    `ctel` varchar(11)NOT NULL COMMENT '手机号',
    `cmoney` int(4)DEFAULT NULL,
    `ostatus` int(2)DEFAULT NULL,
    PRIMARY KEY(`cid`)
)ENGINE=InnoDB AUTO_INCREMENT=9 DEFAULT CHARSET=gbk;
```

2. 家具信息表的建表语句

```
CREATE TABLE `furniture`(
    `fid` int(11)NOT NULL AUTO_INCREMENT,
    `fname` varchar(20)NOT NULL,
    `fsummary` varchar(50)DEFAULT NULL,
    `fnum` int(11)DEFAULT NULL,
    `fmoney` int(4)DEFAULT NULL,
    `fpic` varchar(40)DEFAULT NULL,
    PRIMARY KEY(`fid`)
)ENGINE=InnoDB AUTO_INCREMENT=20 DEFAULT CHARSET=gbk;
```

3. 订单信息表的建表语句

```
CREATE TABLE `oorder`(
    `orderID` int(11)NOT NULL AUTO_INCREMENT,
    `bnumber` int(11)DEFAULT NULL,
    `username` varchar(20)DEFAULT NULL,
    `Truename` varchar(20)DEFAULT NULL,
    `address` varchar(100)DEFAULT NULL,
    `OrderDate` timestamp NULL DEFAULT CURRENT_TIMESTAMP,
    `enforce` int(4)DEFAULT NULL,
    `bz` varchar(200)DEFAULT NULL,
    PRIMARY KEY(`orderID`)
)ENGINE=InnoDB AUTO_INCREMENT=73 DEFAULT CHARSET=gbk;
```

4. 操作员信息表的建表语句

```
CREATE TABLE `operator`(
    `oid` int(11)NOT NULL AUTO_INCREMENT,
    `opassword` varchar(20)DEFAULT NULL,
    `oname` varchar(20)DEFAULT NULL,
```

```
    `ostatus` int(2)DEFAULT NULL,
    PRIMARY KEY(`oid`)
)ENGINE=InnoDB AUTO_INCREMENT=1003 DEFAULT CHARSET=gbk;
```

5．购物车信息表的建表语句

```
CREATE TABLE `order_detail`(
    `ID` int(11)NOT NULL AUTO_INCREMENT,
    `orderID` int(11)DEFAULT NULL,
    `goodsID` int(11)DEFAULT NULL,
    `price` int(11)DEFAULT NULL,
    `number` int(11)DEFAULT NULL,
    PRIMARY KEY(`ID`)
)ENGINE=InnoDB AUTO_INCREMENT=73 DEFAULT CHARSET=gbk;
```

6．公告信息表的建表语句

```
CREATE TABLE `placard`(
    `ID` int(11)NOT NULL AUTO_INCREMENT,
    `title` varchar(100)DEFAULT NULL,
    `content` mediumtext,
    PRIMARY KEY(`ID`)
)ENGINE=InnoDB AUTO_INCREMENT=10 DEFAULT CHARSET=gbk;
```

6.4.2 系统实现总框架简介

在数据库和各种表建立之后,在MyEclipse开发环境中完成实体模型(即JavaBean)部分、控制(即Servlet)部分和视图界面部分。

建立的网站总框架结构如图6.33所示。

从图6.33中可以看到网站实现包括五大部分：src部分、JRE System Library部分、Java EE 5 Libraries部分、Referenced Libraries部分以及WebRoot部分,其中重点在于src和WebRoot两部分。

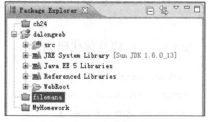

图6.33 网站实现总框架结构图

6.4.3 系统实现——DAL层

在src中包含了四个包：dao包、dao.imp包、Servlet包和vo包,如图6.34所示。

dao包、dao.imp包和vo包构成三层架构中的DAL层；在dao包是定义的接口,在dao.imp包中把dao包中定义的接口给予实现,vo包定义了与数据库中各个表对应的类(即JavaBean),另外定义了名为count的Bean类,以便保存销售统计数据。

1. dao 包

dao 包中定义了实现网站所需的一些接口、DBUtil 类、DAOFactory 类以及由 DAOFactory 派生的 JdbcDAOFactory,如图 6.35 所示。

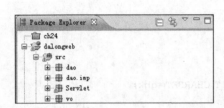
图 6.34　src 的构成

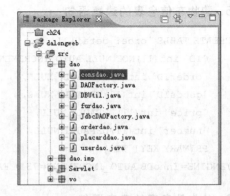
图 6.35　dao 的构成

其中,consdao 是关于普通用户信息类的接口;furdao 是关于家具信息类的接口;orderdao 是关于订单类的接口;placarddao 是关于公告类的接口;userdao 是关于用户信息类的接口。

consdao.java 的具体代码如下:

```
package dao;
import vo.consumer;
public interface consdao{
    public void update(consumer uu,String cname);
    public int getmoney(String username);
    public void updatem(String username,int cmoney);
}
```

furdao.java 的具体代码如下:

```
package dao;
import java.util.List;
import vo.furniture; ;
public interface furdao {
    public furniture getId(int id);
    public void update(furniture f);
    public boolean delete(String id);
    public boolean insert(furniture fur);
    public int getCount(furniture b);
    public int getCount(String sql);
    public List < furniture > getByCondition (furniture b, int fromIndex, int offset);
    public List < furniture > getByCondition (String sql, int fromIndex, int
```

```
        offset);
        public int update(int ID,int num);
}
```

orderdao.java 的具体代码如下：

```
package dao;
import java.util.List;
import vo.order;
import vo.order_detail;
import vo.count;
public interface orderdao {
    public boolean insert(order ord);
    public int getId(order ord);
    public int insert1(order_detail ord1);
    public int getCount(count c,String sdate,String edate);
    public List<count> getByCondition(count c,int fromIndex, int offset,String sdate,String edate);
    public int getSum(String sdate,String edate);
}
```

placarddao.java 的具体代码如下：

```
package dao;
import vo.placard;
public interface placarddao {
    public void insert(placard plac);
    public placard query(int ID);
    public int delete(String bbsID);
}
```

userdao.java 的具体代码如下：

```
package dao;
import java.util.List;
import vo.consumer;
import vo.operator;
public interface userdao {
    public boolean isLogin(String cname,String password,int role);
    public consumer getId(String cname,int role);
    public operator getId1(String id,int role);
    public boolean insert(consumer u);
    public boolean delete(String cname);
}
```

DBUtil 类的作用是建立数据库的驱动和连接，以及相关对象的关闭、查询和更新操作，具体实现代码如下：

```java
package dao;
import java.sql.Connection;
import java.sql.DriverManager;
import java.sql.ResultSet;
import java.sql.SQLException;
import java.sql.Statement;
public class DBUtil {
    private final static String DBDRIVER="com.mysql.jdbc.Driver";
    private final static String DBURL="jdbc:mysql://localhost:3306/dalongweb";
    private final static String DBUSER="root";
    private final static String DBPASSWORD="";
    public static  Connection getConnection(){
        try{
            Class.forName(DBDRIVER);
            Connection conn=DriverManager.getConnection(DBURL, DBUSER, DBPASSWORD);
            return conn;
        }catch(ClassNotFoundException e){
            System.out.println("数据库驱动不存在");
            System.out.println(e.getMessage());
        }catch(SQLException e){
            System.out.println("数据库打开失败");
            System.out.println(e.getMessage());
        }
        return null;
    }
    public static void close(ResultSet rs,Statement st,Connection conn){
        try {
            if(rs!=null)
                rs.close();
            if(st!=null)
                st.close();
            if(conn!=null)
                conn.close();
        } catch(SQLException e){
            e.printStackTrace();
        }
    }
    public ResultSet executeQuery(String sql){
        Connection conn=DBUtil.getConnection();
        ResultSet rs=null;
        Statement st=null;
        try{
```

```java
            st=conn.createStatement();
            rs=st.executeQuery(sql);
            return rs;
        }catch(SQLException ex){
            System.out.println(ex.getMessage());
        }
        return null;
    }
    public int executeUpdate(int ID){
        int result=0;
        String sql="update oorder set enforce=1 where orderID='"+ID+"'";
        Statement stmt=null;
        Connection conn=DBUtil.getConnection();
        try{
            stmt=conn.createStatement();
            stmt.executeUpdate(sql);
                result=1;
            return result;
        }catch(SQLException e){
            e.printStackTrace();
        }finally{
            DBUtil.close(null, stmt, conn);
        }
        return result;
    }
}
```

DAOFactory 类的作用是定义对数据库中表的数据进行增删改的操作，具体代码如下：

```java
package dao;
public abstract class DAOFactory {
    private static DAOFactory factory=new JdbcDAOFactory();
    public static DAOFactory instance(){
        return factory;
    }
    public static DAOFactory instance(String factoryName){
        try{
            Class<?>c=Class.forName(factoryName);
            if(factory.getClass()!=c){
                factory=(DAOFactory)c.newInstance();
            }
            return factory;
        }catch(Exception e){
            e.printStackTrace();
```

```
        }
        return null;
    }
    public abstract userdao getuserdao();
    public abstract furdao getfurdao();
    public abstract orderdao getorderdao();
    public abstract consdao getconsdao();
    public abstract placarddao getplacarddao();
}
```

由于 DAOFactory 是一个抽象类,必须派生子类才能发挥作用,JdbcDAOFactory 就是由 DAOFactory 派生子类,以实现必要的数据访问操作。

```
package dao;
import dao.imp.*;
public class JdbcDAOFactory extends DAOFactory{
    public userdao getuserdao(){
        return new userdaoimpl();
    }
    public consdao getconsdao(){
        return new consdaoimpl();
    }
    public furdao getfurdao(){
        return  new furdaoimpl();
    }
    public orderdao getorderdao(){
        return new orderdaoimpl();
    }
    public placarddao getplacarddao(){
        return new placarddaoimpl();
    }
}
```

2. dao.imp 包

dao.imp 包是对 dao 包中定义的接口进行实现,具体接口的实现如图 6.36 所示。

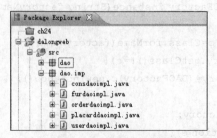

图 6.36　接口实现的各类

其中 consdaoimpl.java 是接口 consdao 的实现，其代码如下：

```java
public class consdaoimpl implements consdao {
    public void update(consumer uu,String cname){
        //TODO Auto-generated method stub
        Connection conn=DBUtil.getConnection();
        String sql="update consumer set cname=?,cpassword=?,ctel=? where cname='"+cname+"'";
        PreparedStatement pst=null;
        try {
            pst=conn.prepareStatement(sql);
            pst.setString(1, uu.getcname());
            pst.setString(2, uu.getcpassword());
            pst.setString(3, uu.getctel());
            pst.executeUpdate();
        } catch(SQLException e){
            e.printStackTrace();
        }finally{
            DBUtil.close(null, null, conn);
        }
    }
    public int getmoney(String username){
        String sql="select cmoney from consumer where cname='"+username+"'";
        Connection conn=null;
        ResultSet rs=null;
        int result=0;
        try{
            conn=DBUtil.getConnection();
            rs=conn.createStatement().executeQuery(sql);
            if(rs.next())
            result=rs.getInt("cmoney");
            return result;
        }catch(SQLException e){
            e.printStackTrace();
        }finally{
            DBUtil.close(rs, null, conn);
        }
        return result;
    }
    public void updatem(String username,int cmoney){
        String sql="update consumer set cmoney='"+cmoney+"' where cname='"+username+"'";
        Connection conn=DBUtil.getConnection();
```

```
        PreparedStatement pst=null;
        try {
            pst=conn.prepareStatement(sql);
            pst.executeUpdate();
        } catch(SQLException e){
            e.printStackTrace();
        }finally{
            DBUtil.close(null, null, conn);
        }
    }
}
```

其他实现代码略。

3. vo 包

vo 包定义了与数据库中各个表对应的 JavaBean 类,类中定义了各项属性、设置器、访问器和构造器,实际上提供了一个公共访问数据的统一接口(以方法方式给出);另外定义了 count 类,以便保存销售统计数据;定义的与表对应的各种 JavaBean 类如图 6.37 所示。

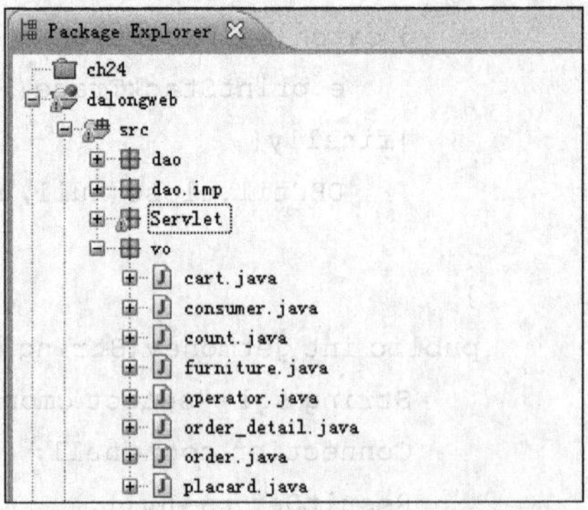

cart.java 中定义了购物车类。具体代码如下:

图 6.37 与表对应的各种类的示图

```
package vo;
public class cart {
    private int ID;
    private int price;
    private int num;
    public void setID(int ID){
        this.ID=ID;
    }
    public int getID(){
        return ID;
    }
    public void setprice(int price){
        this.price=price;
    }
    public int getprice(){
        return price;
    }
    public void setnum(int num){
        this.num=num;
    }
    public int getnum(){
```

```
            return num;
        }
}
```

consumer.java 中定义了用户类。具体代码如下：

```
public class consumer {
    private int cid;
    private String cname;
    private String cpassword;
    private String ctel;
    private int cmoney;
    private int ostatus;
    public void setcid(int cid){
        this.cid=cid;
    }
    public int getcid(){
        return cid;
    }
    public void setcname(String cname){
        this.cname=cname;
    }
    public String getcname(){
        return cname;
    }
    public void setcpassword(String cpassword){
        this.cpassword=cpassword;
    }
    public String getcpassword(){
        return cpassword;
    }
    public void setctel(String ctel){
        this.ctel=ctel;
    }
    public String getctel(){
        return ctel;
    }
    public void setcmoney(int cmoney){
        this.cmoney=cmoney;
    }
    public int getcmoney(){
        return cmoney;
    }
    public void setostatus(int ostatus){
        this.ostatus=ostatus;
```

```
        }
        public int getostatus(){
            return ostatus;
        }
}
```

count.java 中定义了销售类。具体代码如下:

```
package vo;
public class count {
    private String fname;
    private int num;
    private int sum;
    public void setfname(String fname){
        this.fname=fname;
    }
    public String getfname(){
        return fname;
    }
    public void setnum(int num){
        this.num=num;
    }
    public int getnum(){
        return num;
    }
    public void setsum(int sum){
        this.sum=sum;
    }
    public int getsum(){
        return sum;
    }
}
```

furniture.java 中定义了家具类。具体代码如下:

```
package vo;
public class furniture {
    private int fid;
    private String fname;
    private String fsummary;
    private int fnum;
    private int fmoney;
    private String fpic;
    public void setfid(int fid){
        this.fid=fid;
    }
```

```java
    public int getfid(){
        return fid;
    }
    public void setfname(String fname){
        this.fname=fname;
    }
    public String getfname(){
        return fname;
    }
    public void setfsummary(String fsummary){
        this.fsummary=fsummary;
    }
    public String getfsummary(){
        return fsummary;
    }
    public void setfnum(int fnum){
        this.fnum=fnum;
    }
    public int getfnum(){
        return fnum;
    }
    public void setfmoney(int fmoney){
        this.fmoney=fmoney;
    }
    public int getfmoney(){
        return fmoney;
    }
    public void setfpic(String fpic){
        this.fpic=fpic;
    }
    public String getfpic(){
        return fpic;
    }
}
```

operator.java 中定义了操作员类。具体代码如下：

```java
package vo;
public class operator {
    private int oid;
    private String opassword;
    private String oname;
    private int ostatus;
    public void setoid(int oid){
        this.oid=oid;
```

```java
        }
        public int getoid(){
            return oid;
        }
        public void setopassword(String opassword)
        {
            this.opassword=opassword;
        }
        public String getopassword(){
            return opassword;
        }
        public void setoname(String oname){
            this.oname=oname;
        }
        public String getoname(){
            return oname;
        }
        public void setostatus(int ostatus){
            this.ostatus=ostatus;
        }
        public int getostatus(){
            return ostatus;
        }
}
```

order_detail.java 中定义了订单详情类。具体代码如下：

```java
package vo;
public class order_detail {
    private int ID;
    private int orderID;
    private int goodsID;
    private int price;
    private int number;
    public void setID(int ID){
        this.ID=ID;
    }
    public int getID(){
        return ID;
    }
    public void setorderID(int orderID){
        this.orderID=orderID;
    }
    public int getorderID(){
        return orderID;
```

```java
    }
    public void setgoodsID(int goodsID){
        this.goodsID=goodsID;
    }
    public int getgoodsID(){
        return goodsID;
    }
    public void setprice(int price){
        this.price=price;
    }
    public int getprice(){
        return price;
    }
    public void setnumber(int number){
        this.number=number;
    }
    public int getnumber(){
        return number;
    }
}
```

order.java 中定义了订单类。具体代码如下：

```java
package vo;
public class order {
    private int orderID;
    private int bnumber;
    private String username;
    private String Truename;
    private String address;
    private String OrderDate;
    private int enforce;
    private String bz;
    public void setorderID(int orderID){
        this.orderID=orderID;
    }
    public int getorderID(){
        return orderID;
    }
    public void setbnumber(int bnumber){
        this.bnumber=bnumber;
    }
    public int getbnumber(){
        return bnumber;
    }
```

```java
        public void setusername(String username){
            this.username=username;
        }
        public String getusername(){
            return username;
        }
        public void setTruename(String Truename){
            this.Truename=Truename;
        }
        public String getTruename(){
            return Truename;
        }
        public void setaddress(String address){
            this.address=address;
        }
        public String getaddress(){
            return address;
        }
        public void setOrderDate(String OrderDate){
            this.OrderDate=OrderDate;
        }
        public String getOrderDate(){
            return OrderDate;
        }
        public void setenforce(int enforce){
            this.enforce=enforce;
        }
        public int getenforce(){
            return enforce;
        }
        public void setbz(String bz){
            this.bz=bz;
        }
        public String getbz(){
            return bz;
        }
}
```

placard.java 中定义了公告类。具体代码如下：

```java
package vo;
public class placard {
    private int ID;
    private String title;
    private String content;
```

```
        public void setID(int ID){
            this.ID=ID;
        }
        public int getID(){
            return ID;
        }
        public void settitle(String title){
            this.title=title;
        }
        public String gettitle(){
            return title;
        }
        public void setcontent(String content){
            this.content=content;
        }
        public String getcontent(){
            return content;
        }
    }
```

6.4.4 系统实现——USL 层

1. 网站首页部分的设计与实现

网站首页包括五部分,第一部分登录页面,第二部分公告标题显示,第三部分网站标题显示,第四部分家居信息显示,第五部分公司信息显示,具体如图 6.38 所示。

图 6.38 网站首页图

网站首页实现代码在 homepage.jsp 文件中。其中主要代码如下。
1) 获取项目所在目录的代码

```
<%
```

```
String path=request.getContextPath();
String basePath = request.getScheme()+"://"+request.getServerName()+":"+
request.getServerPort()+path+"/";
%>
```

2) 首页布局部分代码

在首页中由五部分组成：第一部分，用户登录部分(login.jsp)；第二部分，公告栏部分(placardmana/placard_show.jsp)；第三部分，网站标题部分(top.jsp)；第四部分，家具信息显示部分(index.jsp)；第五部分，公司联系信息显示(down.jsp)。

网站首页中第一部分实现代码在文件 login.jsp 中，检测输入信息和设置用户动作部分代码如下：

```
<body>
<div id="main">
<table width="100%"  border="0" cellpadding="0" cellspacing="0">
<tr>
<td height="129" align="center" background="images/shop_17.gif">
//对登录事件的响应处理在业务逻辑层中的 LoginServlet.java 中的 case 1 进行
<form name="login" method="POST" action="LoginServlet" onsubmit="return Check();">
<input type="hidden" id="action" name="action" value="1"/>
<table width="128" height="122" style="left: 0px; top: 0px; width: 60px;">
    <%if(request.getAttribute("error")!=null){ %>
    <tr>
    <td align=left colspan=0 height=18>
        <font color=red><b><%=request.getAttribute("error")%></b></font>
    </td>
    </tr>
    <%
        }
    %>
    <tr>
        <td align=left width="80">用户名:</td>
        <td>
        <input type="text" name="name" size="12"></td>
    </tr>
    <tr>
        <td align=left  width="80">密   码:</td>
        <td><input type="password" name="password" size="12"></td>
    </tr>
    <tr>
        <td></td>
        <td align=left>
        <input type="radio" value="0" name="role" id="role" checked="checked"/
        >操作人员
```

```
            <input type="radio" value="1" name="role" id="role" />用户</td>
        </tr>
        <tr>
        <td></td>
        <td align=left>
            <input type="button"name="login"value="登 录" onclick="Check()"/>
```
//对注册事件的响应是执行register.jsp弹出注册页面
```
            <a href="register.jsp " target="middledown">注册</a>
            <script language="javascript">
```
//对用户输入的信息进行检测
```
                function Check()
                {
                    if(document.login.name.value=="")
                    {
                        window.alert("用户名不得为空!");
                        window.login.id.focus();
                        return false;
                    }
                    if(document.login.password.value.length<6)
                    {
                        window.alert("密码长度大于六!");
                        window.login.password.focus();
                        return false;
                    }
                    document.login.submit();
                }
            </script>
        </td>   </tr>   </table>   </form>   </td>   </tr>   </table>
    </div></body></html>
```

网站首页第二部分实现代码在 placard_show.jsp 文件中,代码主要部分如下。
(1) 与数据访问相关的部分代码如下:

```
DBUtil conn=new DBUtil();
ResultSet rs=conn.executeQuery("select  * from placard order by ID desc limit 0,5");
```

(2) 显示公告标题和设置用户动作部分代码如下:

```
<%
    int ID=0;
    String title="";
    while(rs.next()){
    ID=rs.getInt(1);
    title=rs.getString(2);
        %>
```

```
<tr>
    <td height="24" class="tableBorder_T_dashed">
```
//单击某一公告标题产生一事件,该事件在 PlacardServlet.java 中的 case 2 进行处理
```
        <a href="PlacardServlet?action=2&ID=<%=ID %>" target="middledown">
        <%=title%></a>
    </td>
</tr>
<%  }%>
```

网站首页第三部分实现代码在 top.jsp 文件中,其中获取当前时间并显示部分代码如下:

```
<table width="100%" height="20"  border="0" cellpadding="0" cellspacing="0">
<tr align="right"><td></td><td></td><td></td><td></td><td></td>
<%=date.getYear()+1900 %>年
<%=date.getMonth()+1 %>月
<%=date.getDate()%>日
<%=date.getHours()%>时
%=date.getMinutes()%>分</tr>
</table>
```

网站首页第四部分实现代码在 index.jsp 文件中,代码主要部分如下。
(1) 与数据访问相关部分的代码如下:

```
<%String sql="select * from furniture ";
  String d=(String)request.getAttribute("k");
   if(d==null)
   sql="select * from furniture ";
else if(d=="fname")
    sql="select * from furniture order by fname desc";
else if(d=="fsummary")
    sql="select * from furniture order by fsummary desc";
else if(d=="fnum")
    sql="select * from furniture order by fnum desc";
else if(d=="fmoney")
    sql="select * from furniture order by fmoney desc";
    else;
  DBUtil conn=new DBUtil();
  ResultSet rs=conn.executeQuery(sql);
%>
```

(2) 显示商品详细信息和设置用户动作部分代码如下:

```
<body>
```
//按家具名进行查询的事件在 FurServlet.java 中的 case 5 进行处理
```
<form method="post" name="furform" action="FurServlet">
    <input type="hidden" name="action" id="action" value="5"/>
```

```html
<table width="100%" height="23" border="0" cellpadding="0" cellspacing="0">
<tr><td colspan="2" height="7"></td></tr>
<tr>
家具名:<input type="text" id="fname" name="fname" size=66/>
<input type="submit" value="查询"/>
<td bgcolor="#CCCCCC" height="6px"></td>
</tr>
</table>
</form>
```
//设置家具显示项
```html
<table width="100%" height="48" border="1" cellpadding="0" cellpadding="0" bordercolor="FFFFFF">
    <tr>
    <td width="10%" height="24" align="center" onclick="check0(this)">商品号</td><td width="2%"><img src="images/shop_09.gif" width="3" height="57"></td>
    <td width="10%" height="24" align="center" onclick="check1(this)">名称</td>
    <td width="2%"><img src="images/shop_09.gif" width="3" height="57"></td>
    <td width="35%" height="24" align="center" onclick="check2(this)">简介</td>
    <td width="2%"><img src="images/shop_09.gif" width="3" height="57"></td>
    <td width="5%" height="24" align="center" onclick="check3(this)">数量(个)
    </td><td width="2%"><img src="images/shop_09.gif" width="3" height="57">
    </td>
    <td width="5%" height="24" align="center" onclick="check4(this)">价格(元)
    </td><td width="2%"><img src="images/shop_09.gif" width="3" height="57">
    </td>
    <td width="35%" height="24" align="center">图片</td><td width="2%"><img src="images/shop_09.gif" width="3" height="57"></td>
    </tr>
    <%
    String str=(String)request.getParameter("Page");
    if(str==null){
    str="0";
    }
    int pagesize=7;
    rs.last();
    int RecordCount=rs.getRow();
    rs.first();
    int maxpage=0;
    maxpage= (RecordCount%pagesize==0)?(RecordCount/pagesize):(RecordCount/pagesize+1);
    int Page=Integer.parseInt(str);
    if(Page<1){
    Page=1;
    }
    else{
    if(Page>maxpage){
```

```
        Page=maxpage;
    }
   }
   rs.absolute((Page-1)*pagesize+1);
 for(int i=1;i<=pagesize;i++){
   furniture fur=new furniture();
       fur.setfid(rs.getInt("fid"));
       fur.setfname(rs.getString("fname"));
       fur.setfsummary(rs.getString("fsummary"));
       fur.setfnum(rs.getInt("fnum"));
       fur.setfmoney(rs.getInt("fmoney"));
       fur.setfpic(rs.getString("fpic"));
  %>
<tr>
<td height="20" align="center"><%= fur.getfid()%></td><td width="2%"><img src="images/shop_09.gif" width="3" height="50"></td>
<td align="center"><%= fur.getfname()%></td><td width="2%"><img src="images/shop_09.gif" width="3" height="50"></td>
<td align="center"><%= fur.getfsummary()%></td><td width="2%"><img src="images/shop_09.gif" width="3" height="50"></td>
<td align="center"><%= fur.getfnum()%></td><td width="2%"><img src="images/shop_09.gif" width="3" height="50"></td>
<td align="center"><%= fur.getfmoney()%></td><td width="2%"><img src="images/shop_09.gif" width="3" height="50"></td>
<td align=center><img src=<%out.print(fur.getfpic());%> width=80 height=100></td><td width="2%"><img src="images/shop_09.gif" width="3" height="50"></td>
<td align="right">
//购买某一家具事件在FurServlet.java中的case 6进行处理
   <form method="post" name="addfur" action="FurServlet">
   <input type="hidden" name="action" id="action" value=6>
   <input type="hidden" name="id" value=<%=fur.getfid()%>>
   <input type="submit" name="addfur" value="购买"></form>
   </td>
  </tr>
  <%
   try{
   if(!rs.next()){break;}
       }catch(Exception e){}
   }
    %>
 </table>
```

(3) 翻页处理部分的代码如下：

```
<table width="100%" border="0" cellspacing="0" cellpadding="0">
```

```
<tr>
<td align="right">当前页数:[<%= Page %>/<%=maxpage %>] 
<%if(Page>1){ %>
    <a href="index.jsp? Page=1">第一页</a><a href="index.jsp? Page=<%=
    Page-1 %>">上一页</a>
<%}
if(Page<maxpage){ %>
<a href="index.jsp? Page=<%= Page+1 %>">下一页</a><a href="index.jsp? Page=<%=
maxpage %>">最后一页  </a>
<%} %>
</td>
</tr>
</table>
```

网站首页第五部分实现代码见系统中的 down.jsp 文件。

2．注册部分的设计与实现

单击登录页面中的"注册"按钮（即发出注册请求），弹出注册页面，页面如图 6.39 所示。

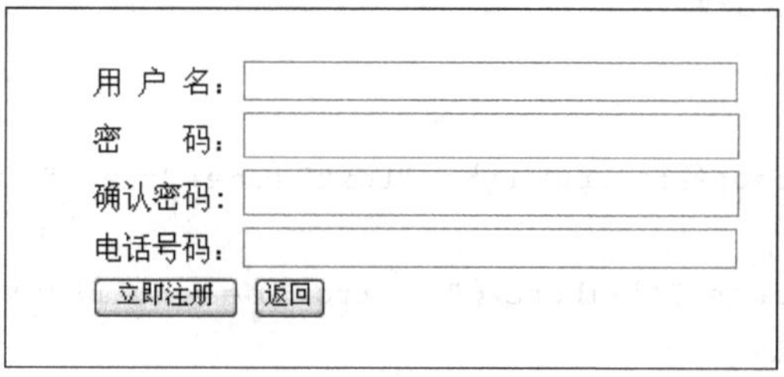

图 6.39　注册页面

在页面中相应区域填写相应信息之后，可以单击"立即注册"按钮，如果填写的信息合理即可完成注册；如果单击"返回"按钮，即返回到上一页面。对应的实现代码在文件 register.jsp 中。检测输入信息和设置用户动作部分的代码：

```
<script type="text/javascript">
 function check(){
 if(document.register.cname.value=="")
 {
window.alert("用户名不为空!");
window.register.cname.focus();
return false;
}
if(document.register.password.value=="")
{
window.alert("请输入密码!");
window.register.password.focus();
return false;
```

```
            }
            if(document.register.password.value!=document.register.confpwd.value)
            {
            window.alert("两次输入密码不一致,请重新输入");
            window.register.confpwd.focus();
            return false;
            }
            document.register.submit();
            }
            </script>
        </head>
        <body>
        <div id="dv">用户注册</div>
        <hr/>
        <br/>
        //单击"立即注册"按钮产生的事件在 LoginServlet.java 中的 case 4 进行处理
        <form name="register" action="LoginServlet " method="post">
        <input type="hidden" name="action" value="4"/>
        <table align="center">
        <tr>
        <td>
            用 户 名:<input type="text" name="cname" size=36 />
        </td>
        <td><%=request.getAttribute("iderror")==null?"":request.getAttribute("iderror")%></td>
        </tr>
        <tr>
        <td>密    码:<input type="password" name="password" size=40/>
        </td>
        </tr>
        <tr><td>
            确认密码: <input type="password" name="confpwd" size=40/>
        </td></tr>
        <tr><td>
            电话号码:<input type="text" name="tel" size=36 />
        </td></tr>
        <tr>
        <td><input type="button" value="立即注册" onclick="check()">
        //单击"返回"按钮即返回到上一页面
        <input type="button" value="返回" onclick="JScript:history.back(-1)"></td>
        </tr>
        </table>
        </form>
```

3．操作员功能部分的实现

1）功能菜单页面的设计与实现

在登录页面中，以操作员身份登录后，进入操作员功能菜单页面，如图6.40所示。

操作员功能菜单页面实现代码在文件 oper.jsp 中。其中功能菜单的显示和设置用户动作部分代码如下。

图6.40 操作员功能菜单页面

```
<table width="92%" height="48" border="0" align=
"center" cellpadding="0" cellspacing="0">
```
//单击商品管理条目产生的事件在 OperServlet.java 中的 case 1 进行处理
```
        <tr><a href="OperServlet?action=1" target="middledown">商品管理</a>
        </tr>
```
//单击销售统计条目产生的事件执行 furmana/furcount.jsp，弹出销售统计页面
```
        <tr><a href="furmana/furcount.jsp" target="middledown">销售统计</a></tr>
```
//单击会员管理条目产生的事件执行 operatormana/membermana.jsp，弹出会员管理页面
```
        <tr><a href="operatormana/membermana.jsp" target="middledown">会员管
理</a></tr>
```
//单击公告管理条目产生的事件在 OperServlet.java 中的 case 5 进行处理
```
        <tr><a href="OperServlet?action=5" target="middledown">公告管理</a>
        </tr>
```
//单击商品管理条目产生的事件在 OperServlet.java 中的 case 3 进行处理
```
        <tr><a href="OperServlet?action=3" target="middledown">订单管理</a>
        </tr>
```
//单击商品管理条目产生的事件在 OperServlet.java 中的 case 4 进行处理
```
        <tr><a href="OperServlet?action=4" target="_parent">注销</a></tr>
</table>
```

2）商品管理功能页面的设计与实现

在操作员功能菜单页面中选中商品管理功能后（即发出商品管理请求），进入商品管理页面，页面如图6.41所示。

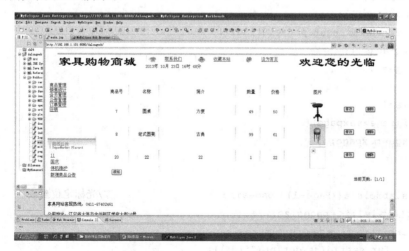

图6.41 操作员商品管理页面

商品管理页面的实现代码在文件 furmana.jsp 中,其中显示商品详细信息和设置用户动作部分代码如下:

```jsp
<body>
        <table width="100%" height="48" border="1" cellpadding="0" cellpadding="0"
            bordercolor="FFFFFF">
    <tr>
    <td width="6%" height="24" align="center">商品号</td><td width="2%"><img src="images/shop_09.gif" width="3" height="57"></td>
    <td width="6%" height="24" align="center">名称</td><td width="2%"><img src="images/shop_09.gif" width="3" height="57"></td>
    <td width="20%" height="24" align="center">简介</td><td width="2%"><img src="images/shop_09.gif" width="3" height="57"></td>
    <td width="5%" height="24" align="center">数量</td><td width="2%"><img src="images/shop_09.gif" width="3" height="57"></td>
    <td width="5%" height="24" align="center">价格</td><td width="2%"><img src="images/shop_09.gif" width="3" height="57"></td>
    <td width="15%" height="24" align="center">图片</td><td width="2%">
    </tr>
    <%
    String str=(String)request.getParameter("Page");
    if(str==null){
    str="0";
    }
    int pagesize=7;
    rs.last();
    int RecordCount=rs.getRow();
    rs.first();
    int maxpage=0;
    maxpage= (RecordCount%pagesize==0)?(RecordCount/pagesize):(RecordCount/pagesize+1);
    int Page=Integer.parseInt(str);
    if(Page<1){
    Page=1;
    }
    else{
    if(Page>maxpage){
    Page=maxpage;
    }
    }
    rs.absolute((Page-1)*pagesize+1);             //光标定位到该行
   for(int i=1;i<=pagesize;i++){
    furniture fur=new furniture();
        fur.setfid(rs.getInt("fid"));
```

```jsp
        fur.setfname(rs.getString("fname"));
        fur.setfsummary(rs.getString("fsummary"));
        fur.setfnum(rs.getInt("fnum"));
        fur.setfmoney(rs.getInt("fmoney"));
        fur.setfpic(rs.getString("fpic"));
     %>
     <tr>
     <td height="20" align="center"><%=fur.getfid()%></td><td width="2%"><img src="images/shop_09.gif" width="3" height="57"></td>
     <td align="center"><%=fur.getfname()%></td><td width="2%"><img src="images/shop_09.gif" width="3" height="57"></td>
     <td align="center"><%=fur.getfsummary()%></td><td width="2%"><img src="images/shop_09.gif" width="3" height="57"></td>
     <td align="center"><%=fur.getfnum()%></td><td width="2%"><img src="images/shop_09.gif" width="3" height="57"></td>
     <td align="center"><%=fur.getfmoney()%></td><td width="2%"><img src="images/shop_09.gif" width="3" height="57"></td>
     <td align="center"><img src=<%out.print(fur.getfpic());%> width=60 height=80></td>
     <td align="left" width="6%">
//单击"修改"按钮产生的事件在FurServlet中的case 1进行处理
     <form method="post" name="changeform" action="FurServlet">
     <input type="hidden" name="action" id="action" value="1"/>
     <input type="hidden" name=id value=<%=fur.getfid()%>>
     <input type="submit" name="change"  value="修改">
     </form></td>
     <td align="left"  width="6%">
//单击"删除"按钮产生的事件在FurServlet中的case 3进行处理

     <form method="post" name="deleteform" action="FurServlet">
     <input type="hidden" name="action" id="action" value=3>
     <input type="hidden" name="id" value=<%=fur.getfid()%>>
     <input type="submit" name="delete" value="删除"></form>
     </td>    </tr>
     <%
      try{
      if(!rs.next()){break;}
         }catch(Exception e){}
       }
        %>
//单击"添加"按钮产生事件执行furmana/furadd.jsp,弹出添加商品页面
     <td align="center"><input type="button" value="添加" onclick="window.location.href='furmana/furadd.jsp'"/></td>
     </table>
```

```
<table width="100%" border="0" cellspacing="0" cellpadding="0">
    <tr>
```

3) 修改商品页面的设计与实现

在商品管理页面中对某一商品进行修改(即发出修改请求),进入商品修改页面,如图 6.42 所示。

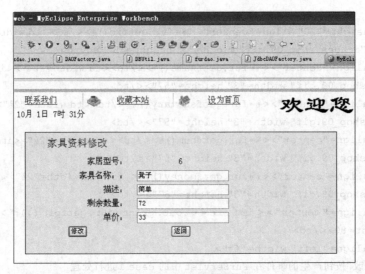

图 6.42 修改商品页面

修改商品页面实现代码在文件 furchange.jsp 中,其中显示商品信息和设置用户动作部分代码如下:

```
<body>
//单击"修改"按钮产生 FurServlet.java 中的 case 2 进行处理
<form method="post" name="fchangeform" action="FurServlet">
<%
        furniture f=(furniture)request.getAttribute("furchange");
%>
        <input type="hidden" name="action" value="2"/>
            <input type="hidden" name="fid" value="<%=f.getfid()%>"/>
<table align="center" bgcolor="#EEEEEE" style="border:1px black solid">
    <tr>
        <td colspan="3" align="center" style="text-align:center;height:40px;
font-size:20px">家具资料修改</td>
    </tr><tr>
        <td align="right">家居型号:</td>
        <td align="center"><%=f.getfid()%></td>
        <td></td></tr>
    <tr>
        <td align="right">家具名称::</td>
        <td align="center">< input type="text" name="fname" value="<%=
```

```
                    f.getfname()%>" class="txt"/></td>
                    <td></td></tr>
        <tr>
                    <td align="right">描述:</td>
                    <td align="center"><input type="text" name="fsummary" value=
                    "<%= f.getfsummary()%>" class="txt"/></td>
                    <td></td>    </tr>
        <tr>
                    <td align="right">剩余数量:</td>
                    <td align="center">< input type="text" name="fnum" value="<%=
                    f.getfnum()%>" class="txt"/></td>
                    <td></td>           </tr>           <tr>
                    <td align="right">单价:</td>
                    <td align="center">< input type="text" name="fmoney" value="<%=
                    f.getfmoney()%>" class="txt"/></td>
                    <td></td></tr>
<tr>
                    <td align="center"><input type="submit" value="修改"/></td>
//单击"返回"按钮则返回到上一页面
                    <td align="center"><input type="button" value="返回" onclick=
                    "JScript:history.back(-1)"/></td>
                    <td></td></tr><tr><td colspan="3" height="40px"></td></tr>
                        </table>
</form>
</body>
```

4）添加商品页面的设计与实现

在商品管理页面中选择添加商品（即发出添加请求），进入商品添加页面，如图 6.43 所示。

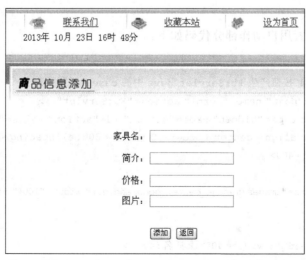

图 6.43　商品添加页面

实现添加商品页面的代码在文件 furadd.jsp 中。

对输入数据进行检测部分代码如下：

```
<script language="javascript">
function check(){
    if(document.form1.fname.value==""){
        window.alert("家具名不为空!");
    window.furadd.fname.focus();
    return false;
     }
     if(document.form1.fsummary.value==""){
        window.alert("请输入商品简介!");
    window.furadd.fsummary.focus();
    return false;
     }
     if(isNaN(form1.fmoney.value))
    {
        alert("商品定价错误!");
        document.form1.fmoney.value="";
        document.form1.fmoney.focus();
        return false;
    }
    if(document.form1.fpic.value=="")
    {
        alert("请输入图片文件的路径!");
        document.form1.fpic.focus();
        return false;
    }
    form1.submit();
    }
```

显示信息和设置用户动作部分代码如下：

```
<body>
    //单击"添加"按钮产生 FurServlet.java 中的 case 4 进行处理
<form method="post" name="form1" action="FurServlet"  >
    <input type="hidden" name="action" id="action" value="4"/>
    <table align = center border = 0 width = 80% cellspacing = 2 cellpadding = 0 height=400>
    <tr>
    <img src="images/manage_center_goodsadd.gif" width="100%" height="128">
</tr>
<tr>
<td align=right width=40%>家具名:</td>
<td colspan=2 width=70%><input type="text" name="fname"></td>
```

```
            </tr>
            <tr>
                <td align=right width=40%>简介:</td>
                <td colspan=2 width=70%><input type="text" name="fsummary"></td>
            </tr>
            <tr>
                <td align=right width=40%>价格:</td>
                <td colspan=2 width=70%><input type="text" name="fmoney"></td>
            </tr>
            <tr>
                <td align=right width=40%>图片:</td>
                <td colspan=2 width=70%><input type="text" name="fpic"></td>
            </tr>
            <tr>
                <td align=left> </td>
                <td align=left colspan=2><input type=button value="添加" onclick="check()">
                //单击"返回"按钮则返回到上一页面
                <input type="button" value="返回" onclick="JScript:history.back(-1)"/>
                </td>
            </tr>
        </table>
    </form>
</body>
```

5) 销售统计页面的设计与实现

在操作员功能菜单中选择销售统计(即发出销售统计请求),进入销售统计页面,如图 6.44 所示。

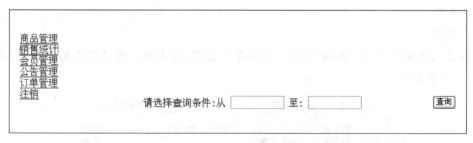

图 6.44　销售统计页面

销售统计页面实现代码在文件 furcount.jsp 中。

对输入数据进行检测部分代码如下:

```
<script language="javascript">
    function check(){
    if(form1.sdate.value==0){
        form1.sdate.focus();
        alert("请输入查询日期");return false;
```

```
            }
    if(form1.edate.value==0){
    form1.edate.focus();
    alert("请输入截止日期");return false;
        }
form1.submit();
        }
</script>
</head>
```

显示信息和设置用户动作部分代码如下。

```
<body>
//单击"查询"按钮产生的事件在OrdServlet.java中的case 2进行处理
<form method="post" name="form1" action="OrdServlet"  >
    <input type="hidden" name="action" id="action" value="2"/>
    <table width="100%"  border=0 align="center" cellpadding="-2" cellspacing="-2"
        bordercolordark="#FFFFFF">
  ⋮
<td align="center">
            请选择查询条件:从
<input type="text" name="sdate" value="" size="11">
            至：
<input type="text" name="edate" value="" size="11">
</td>
<td width=200px>
<input type=button value="查询" onclick="check()">
  ⋮
</form>
</body>
```

在输入条件后,单击"查询"按钮。如果条件合理,进入销售统计结果显示页面,页面如图6.45所示。

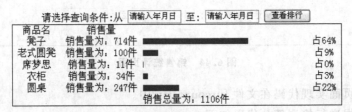

图6.45 销售统计结果显示页面

6) 会员管理页面的设计与实现

在操作员功能菜单中选择会员管理(即发出会员管理请求),进入会员管理页面,如

图 6.46 所示。

图 6.46 会员管理页面

会员管理页面实现代码在文件 membermana.jsp 中。主要代码如下。
(1) 与数据库操作相关部分代码如下：

```
<%DBUtil conn=new DBUtil();
    ResultSet rs=conn.executeQuery("select * from consumer ");
%>
<%
String path=request.getContextPath();
String basePath = request.getScheme()+"://"+ request.getServerName()+":"+
request.getServerPort()+path+"/";
%>
```

(2) 显示信息和设置用户动作部分代码如下：

```
<body>
  ⋮
<tr>
        <td width="50" height="27" align="center">用户名</td>
        <td width="150" align="center">密码</td>
        <td width="150" align="center">电话</td>
        <td width="100" align="center">金额</td>
</tr>
<%
    String str=(String)request.getParameter("Page");
    if(str==null){
    str="0";
    }
    int pagesize=7;
    rs.last();
    int RecordCount=rs.getRow();
    rs.first();
    int maxpage=0;
```

```
maxpage= (RecordCount%pagesize==0)?(RecordCount/pagesize):(RecordCount/pagesize+1);
int Page=Integer.parseInt(str);
if(Page<1){
Page=1;
}
else{
if(Page>maxpage){
Page=maxpage;
}
}
rs.absolute((Page-1)*pagesize+1);
for(int i=1;i<=pagesize;i++){
consumer con=new consumer();
con.setcname(rs.getString("cname"));
con.setcpassword(rs.getString("cpassword"));
con.setctel(rs.getString("ctel"));
con.setcmoney(rs.getInt("cmoney"));
%>
<tr style="padding:5px;">
<td align="center"><%=con.getcname()%></td>
<td align="center"><%=con.getcpassword()%></td>
<td align="center"><%=con.getctel()%></td>
<td align="center"><%=con.getcmoney()%></td>
<td align="left"  width="6%">
//单击"删除"按钮产生的事件在OperServlet.java中的case 6进行处理
<form method="post" name="deleteform" action="OperServlet">
<input type="hidden" name="action" id="action" value=6>
<input type="hidden" name="cname" value=<%=con.getcname()%>>
<input type="submit" name="delete" value="删除"></form>
</td>
</tr>
<%
 try{
 if(!rs.next()){break;}
   }catch(Exception e){}
 }
  %>
</table>
```

翻页部分代码略。

在操作员功能菜单中选择公告管理(即发出公告管理请求),进入公告管理页面,如图 6.47 所示。

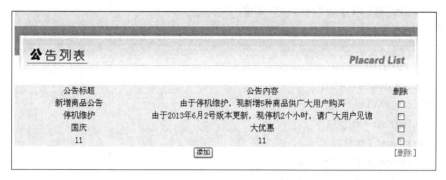

图 6.47 公告管理页面

公告管理页面实现的代码在文件 placard_mana.jsp 中。主要代码如下。
(1) 与数据库操作相关部分代码如下：

```
<%
    DBUtil conn=new DBUtil();
    ResultSet rs=conn.executeQuery("select　*　from placard ");
%>
<%
String path=request.getContextPath();
String basePath = request.getScheme () +"://" + request.getServerName () +":" +
request.getServerPort()+path+"/";
%>
```

(2) 信息显示与设置用户动作部分代码如下：

```
<script type="text/javascript">
    function checkdel(){
    if(confirm("确定要删除吗？")){
    document.placardmanaform.submit();
    }
    }
</script>
<body>
//单击"删除"按钮产生的事件在 PlacardServlet 中的 case 3 进行处理
<form method="post" name="placardmanaform" action="PlacardServlet">
<input type="hidden" name="action" value="3"/>
<table width="100%" border="0" cellspacing="0" cellpadding="0">
<tr>
<img src="images/manage_center_placardlist.gif" width="100%" height="156">
</tr>
<tr bgcolor="#eeeeee">
<td width="32%" height="24" align="center">公告标题</td>
<td width="60%" align="center">公告内容</td>
```

```
            <td width="8%" align="center">删除</td>
        </tr>
        <%
            int ID=0;
            String content="";
            String title="";
            while(rs.next()){
            ID=rs.getInt(1);
            title=rs.getString(2);
            content=rs.getString(3);
             %>
        <tr bgcolor="#eeeeee">
            <td width="32%" height="24" align="center"><%=title %></td>
            <td width="60%" align="center"><%=content %></td>
            <td align="center"><input name="delid" type="checkbox" class="noborder"
            value="<%=ID %>"></td>
        </tr>
        <%} %>
        </table>
        <table width="100%" border="0" cellspacing="0" cellpadding="0">
        <tr>
        //单击"添加"按钮执行placardmana/placard.jsp,弹出添加页面
            <td></td><td width="8%" height="24" align="right"><input type="button"
            name=" addplacard" value=" 添加" onclick=" window. location. href =
            'placardmana/placard.jsp'"></td>
            <td width="8%" align="right">
            [<a style="color:red;cursor:hand;" onclick="checkdel()">删除</a>]
            <div id="ch"><input name="delid" type="checkbox" class="noborder" value=
            "0"></div>
            <script language="javascript">ch.style.display="none";</script>
            </td>
        </tr>
        </table>
        </form>
        </body>
```

在公告管理页面中,选择某些公告进行删除(即发出删除请求),弹出确定窗口如图 6.48 所示。

在公告管理页面单击"添加"按钮(即发出添加请求)弹出添加公告页面,如图 6.49 所示。

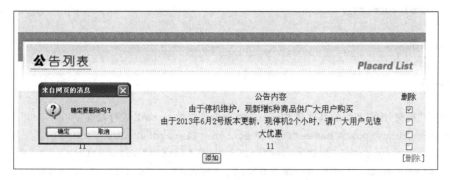

图 6.48　单击删除某一公告弹出的确定对话框

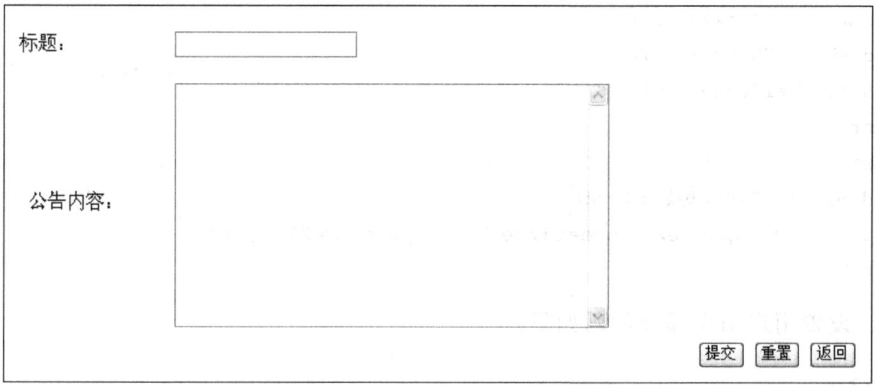

图 6.49　添加公告页面

添加公告的实现代码在文件 placard.jsp 中。主要代码如下。
(1) 与数据库操作相关部分代码如下：

```
<%@page language="java" import="java.util.*" pageEncoding="gb2312"%>
<%
String path=request.getContextPath();
String basePath = request.getScheme()+"://"+ request.getServerName()+":"+
request.getServerPort()+path+"/";
%>
```

(2) 输入信息检测部分代码如下：

```
<script language="javascript">
function mycheck(){
    if(form1.title.value==""){
        alert("请输入公告标题!");form1.title.focus();return;
    }
    if(form1.content.value==""){
        alert("请输入公告内容!");form1.content.focus();return;
    }
```

```
            form1.submit();
        }
        </script>
        </head>
        <body>
        ⋮
//单击"提交"按钮产生的事件在 PlacardServlet 中的 case 1 进行处理
<form method="post" name="form1" action="PlacardServlet" >
    <input type="hidden" name="action" id="action" value="1"/>
    <table width="100%"  border=0 align="center" cellpadding="-2" cellspacing="-2"
    bordercolordark="#FFFFFF">
<tr height=100px>
<td width=140px></td>
<td width=200px></td>
<td width=200px></td>
</tr>
<tr>
<td height="45">标题:</td>
<td><input type=text name=title id=title size=27></td>
</tr>
```

(3) 设置用户动作部分代码如下:

```
<tr>
<td width="14%" height="272" rowspan="2"> 公告内容:</td>
<td width="86%"><textarea name="content" id="content" cols="60" rows="15"
class="textarea"></textarea></td>
</tr>
<tr>
<td width=140px></td>
<td width=200px>
<input type=button value="提交" onclick="mycheck()">
//单击"重置"按钮则清除填写的内容
<input type=reset value="重置">
//单击"返回"按钮则返回上一页面
<input type=button value="返回" onclick="JScript:history.back(-1)">
</td>
<td width=200px></td>
</tr>
    </table>
</form>
</body>
```

操作员功能菜单页面中单击"订单管理"(即发出订单管理请求),进入订单管理页面,页面如图 6.50 所示。

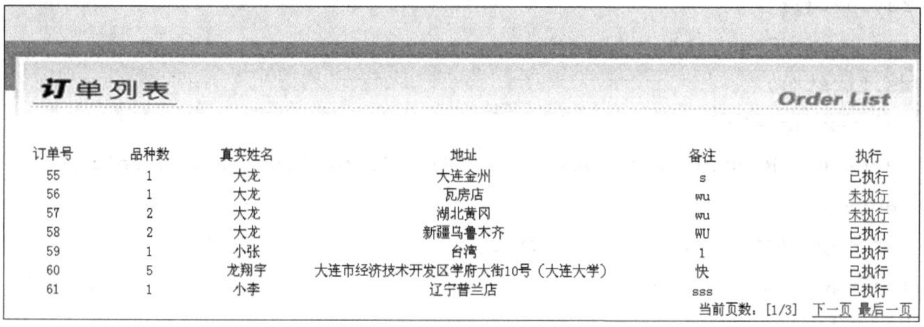

订单号	品种数	真实姓名	地址	备注	执行
55	1	大龙	大连金州	s	已执行
56	1	大龙	瓦房店	wu	未执行
57	2	大龙	湖北黄冈	wu	未执行
58	2	大龙	新疆乌鲁木齐	WU	已执行
59	1	小张	台湾	1	已执行
60	5	龙翔宇	大连市经济技术开发区学府大街10号（大连大学）	快	已执行
61	1	小李	辽宁普兰店	sss	已执行

当前页数：[1/3] 下一页 最后一页

图 6.50 订单管理页面

订单管理页面实现代码在文件 ordermanage.jsp 中。主要代码如下。

(1) 与数据库操作相关部分代码如下：

```
<%DBUtil conn=new DBUtil();
  ResultSet rs=conn.executeQuery("select * from oorder ");
%>
<%
String path=request.getContextPath();
String basePath = request. getScheme ( ) +"://" + request. getServerName ( ) +":" +
request.getServerPort()+path+"/";
%>
```

(2) 显示信息和设置用户动作部分代码如下：

```
<body>
<table width="100%" height="48" border="1" cellpadding="0" cellpadding="0"
bordercolor="FFFFFF">
<tr>
<img src="images/manage_center_orderlist.gif" width="100%" height="156">
</tr>
    <tr>
    <td width="10%" height="24" align="center">订单号</td>
    <td width="10%" height="24" align="center">品种数</td>
    <td width="10%" height="24" align="center">真实姓名</td>
    <td width="35%" height="24" align="center">地址</td>
    <td width="25%" height="24" align="center">备注</td>
    <td width="10%" height="24" align="center">执行</td>
    </tr>
<%
    String str=(String)request.getParameter("Page");
    if(str==null){
    str="0";
    }
    int pagesize=7;
```

```
                rs.last();
                int RecordCount=rs.getRow();
                rs.first();
                int maxpage=0;
                maxpage=(RecordCount%pagesize==0)?(RecordCount/pagesize):(RecordCount/
                pagesize+1);
                int Page=Integer.parseInt(str);
                if(Page<1){
                Page=1;
                }
                else{
                if(Page>maxpage){
                Page=maxpage;
                }
                }
                rs.absolute((Page-1)*pagesize+1);
            for(int i=1;i<=pagesize;i++){
                order ord=new order();
                ord.setorderID(rs.getInt("orderID"));
                ord.setbnumber(rs.getInt("bnumber"));
                ord.setTruename(rs.getString("Truename"));
                ord.setaddress(rs.getString("address"));
                ord.setbz(rs.getString("bz"));
                ord.setenforce(rs.getInt("enforce"));
                %>
<tr>
    <td height="20" align="center"><%=ord.getorderID()%></td>
    <td align="center"><%=ord.getbnumber()%></td>
    <td align="center"><%=ord.getTruename()%></td>
    <td align="center"><%=ord.getaddress()%></td>
    <td align="center"><%=ord.getbz()%></td>
    <td align="center"><%if(ord.getenforce()==0){%>
//单击未执行产生的事件是执行 ordermana/orderenforce.jsp,弹出响应页面(执行成功
与否)
    <a href="ordermana/orderenforce.jsp?ID=<%=ord.getorderID()%>">未执行</a>
    <%}else{ %>已执行<% } %></td>
</tr>
<%
    try{
    if(!rs.next()){break;}
        }catch(Exception e){}
    }
    %>
```

```
            </table>
```

翻页部分代码略。

在订单管理页面中选中某一未执行的订单,单击未执行(即发出未执行请求),则执行该订单,其代码如下:

```
orderenforce.jsp
  ⋮
<%
DBUtil conn=new DBUtil();
if(request.getParameter("ID")!=""){
int ID=Integer.parseInt(request.getParameter("ID"));
int ret=0;
ret=conn.executeUpdate(ID);
if(ret!=0){
out.println("<script language='javascript'>alert('订单执行成功!');window.
location.href='ordermanage.jsp';</script>");
}else{
out.println("<script language='javascript'>laert('订单执行失败!');window.
location.href='ordermanage.jsp';</script>");
}
}
%>
```

4. 以用户身份登录部分

在登录页面中,以用户身份登录系统,进入用户功能菜单页面,如图 6.51 所示。

用户功能菜单页面实现代码在文件 cons.jsp 中。

其中用户功能菜单实现部分代码如下:

图 6.51　用户功能菜单页面

```html
<body>
    <table width="92%" height="48" border="0" align="center" cellpadding="0"
    cellspacing="0">
        <tr><a href="ConsServlet?action=1" target="middledown">个人信息修改
        </a></tr>
        <tr><a href="ConsServlet?action=4" target="middledown">购物车</a>
        </tr>
        <tr><a href="ConsServlet?action=5" target="middledown">订单查询</a>
        </tr>
        <tr><a href="ConsServlet?action=3" target="_parent">注销</a></tr>
    </table>
</body>
```

在用户功能菜单页面中单击修改个人信息(即发出修改个人信息请求),进入修改个人信息页面,如图 6.52 所示。

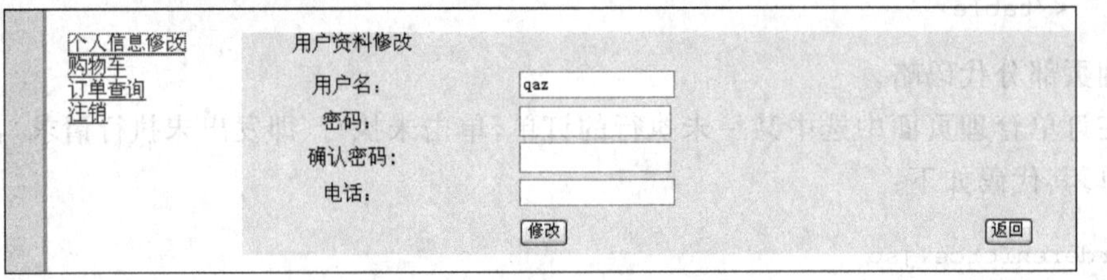

图 6.52 修改个人信息页面

修改个人信息页面实现代码在文件 changecons.jsp 中。主要代码如下。
(1) 对输入数据进行检测部分代码如下：

```
<script type="text/javascript">
    function check(){
    if(document.changeform.cname.value=="")
    {
    window.alert("用户名不为空!");
    window.changeform.cname.focus();
    return false;
    }
    if(document.changeform.password.value=="")
    {
    window.alert("请输入密码!");
    window.changeform.password.focus();
    return false;
    }
    if(document.changeform.password.value!=document.changeform.confpwd.value)
    {
    window.alert("两次输入密码不一致,请重新输入");
    window.changeform.confpwd.focus();
    return false;
    }
    document.changeform.submit();
    }
</script>
```

(2) 显示信息和设置用户动作部分代码如下：

```
<body>
<form method="post" name="changeform" action="ConsServlet">
    <input type="hidden" name="action" value="2"/>
    <input type="hidden" name="id" value="<%=uname %>"/>
    <table width="100%"  border="0" cellspacing="-2" cellpadding="-2" bgcolor=
    "#EEEEEE">
    <tr>
        <td valign="top"  align="center" height="28">用户资料修改</td>
```

```html
            </tr>
            <tr>
                <td height="28" align="center">用户名:</td>
                <td height="28"><input type="text" name="cname" value="<%=uname %>"
                    class="txt" size=16></td>
    <td><%=request.getAttribute("iderror")==null?"":request.getAttribute
    ("iderror")%></td>
            </tr>
            <tr>
                <td height="28" align="center">密码:</td>
                <td height="28"><input type="password" name="password" size=16/></td>
            </tr>
            <tr>
                <td height="28" align="center">确认密码:</td>
                <td height="28"><input type="password" name="confpwd" size=16 /></td>

            </tr>
            <tr>
                    <td height="28" align="center">电话:</td>
                    <td height="28"><input type="text" name="ctel" class="txt" size=
                    16 /></td>
            </tr>
            <tr>
                <td height="34"> </td>
                    <td class="word_grey"><input type="button" value="修改" onclick=
                    "check()"/></td>
                    <td class="word_grey"><input type="button" value="返回" onclick=
                    "JScript:history.back(-1)"/></td>
            </tr>
        </table>
    </form>
    </body>
```

在用户功能菜单页面中单击购物车(即发出购物车请求),进入购物车页面,购物车空时显示的页面如图 6.53 所示,购物车非空时显示的页面如图 6.54 所示。

购物车页面实现代码在文件 cart-see.jsp 中。

其中显示购物车中的信息和设置用户动作部分代码如下:

```html
<body>
<form method="post" name="form1" action="cart/cart-modify.jsp">
    <table width="92%" height="48" border="0" align="center" cellpadding="0"
    cellspacing="0">
        <tr align="center" valign="middle">
            <td height="27" class="tableBoarder_B1">编号</td>
```

```jsp
            <td height="27" class="tableBoarder_B1">商品编号</td>
            <td   class="tableBoarder_B1">商品名称</td>
            <td height="27" class="tableBoarder_B1">单价</td>
            <td height="27" class="tableBoarder_B1">数量</td>
            <td height="27" class="tableBoarder_B1">金额</td>
            <td   class="tableBoarder_B1">退回</td>
    </tr>
<%String username="";
username=(String)session.getAttribute("consumer");
if(username==""||username==null){
out.println("<script language='javascript'>alert('请先登录!');window.
location.href='index1.jsp';</script>");
}else{
DBUtil conn=new DBUtil();
Vector ct=(Vector)session.getAttribute("ct");    //自动增长的对象数组
if(ct==null||ct.size()==0){
out.println("<script language='javascript'>window.location.href=
'cart/cart-null.jsp';</script>");
}else{
        int sum=0;
int ID=-1;
int snum=0;                                   //待修改数量
int finum=0;                                  //该商品总量
String goodsname="";
for(int i=0,j=0,k=0;i<ct.size();i++)
{
snum=i;
    cart cb=(cart)ct.elementAt(i);
    j=cb.getnum();
    k=cb.getprice();
    //sum=sum+j*k;
    ID=cb.getID();
    if(ID>0){
ResultSet rs=conn.executeQuery("select * from furniture where fid=
"+ID);
    if(rs.next()){
        goodsname=rs.getString("fname");
    }
    finum=rs.getInt("fnum");
    if(j<=finum){sum=sum+j*k;}
    else { sum=sum+finum*k; }
}%>
        <tr align="center" valign="middle">
            <td width="32" height="27"><%=i+1 %></td>
```

```jsp
<td width="109" height="27"><%=ID %></td>
<td width="199" height="27"><%=goodsname%></td>
<td width="59" height="27"><%=k %></td>
<td width="51" height="27">
<%if(j<=finum){%>
<input name="num<%=i %>" size="7" type="text" class="txt_grey" value="<%=j %>" onBlur="check(this.form)">
<%}else{ %>
<input name="num<%=i %>" size="7" type="text" class="txt_grey" value="<%=finum %>" onBlur="check(this.form)">
<%} %>
</td>
<td width="65" height="27">$<%=k*j %></td>
<td width="34"><a href="cart/cart-move.jsp?ID=<%=i %>">移除</a></td>
<script language="javascript">
function check(myform){

    if(isNaN(myform.num<%=i%>.value)||myform.num<%=i%>.value.indexOf('.',0)!=-1){
    alert("请不要输入非法字符");myform.num<%=i%>.focus();
    return;
    }
    if(myform.num<%=i%>.value==""){
    alert("请输入修改的数量");myform.num<%=i%>.focus();return;
    }
    myform.submit();
    }
</script>
<%}%>
</tr></table>
<table width="92%" height="48" border="0" align="center" cellpadding="0" cellspacing="0">
<tr align="right" valign="middle">
    <td width="65" height="27">$<%=sum %></td>
</tr>
<tr>
<a href="index.jsp">继续购物</a>
<a href="cart/cart-clear.jsp">清空购物车</a>
<a href="cart/cart-modify.jsp?num=<%= snum %>">修改数量</a>
<a href="ordermana/order.jsp">去购物台结账</a>
</tr>
</table>
<%
```

```
            }%>
    </form>
    </body>
    </html>
```

图 6.53 购物车空时显示的页面

图 6.54 购物车非空时显示的页面

在购物车非空时显示的页面中单击清空购物车(即发出清空购物车请求),把购物车清空,返回如图 6.53 所示的界面。执行的代码是 cart-null.jsp。

删除购物车中物品的实现代码在文件 cart-see1.jsp 中。

其中删除购物车中数据部分代码如下:

```
<%
    Vector ct=(Vector)session.getAttribute("ct");
    try{
    int id=Integer.parseInt(request.getParameter("ID"));
    ct.removeElementAt(id);
    session.setAttribute("ct",ct);
    response.sendRedirect("cart-see1.jsp");
    }catch(Exception e){}
%>
```

单击购物车页面中的去购物台结账(即发出去购物台结账请求),进入结账页面,如图 6.55 所示。

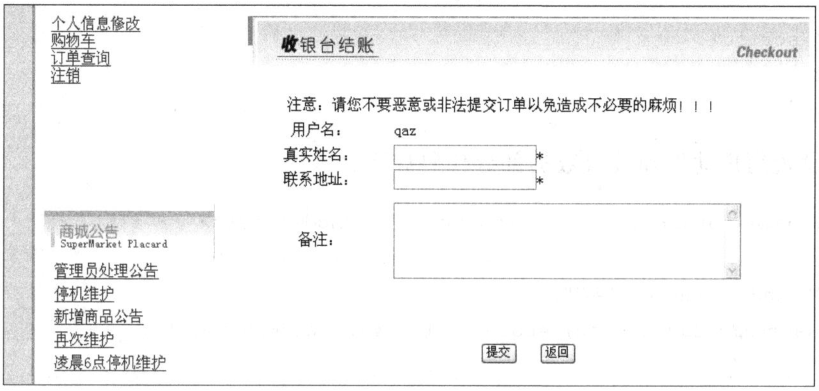

图 6.55 购物结账页面

购物结账页面实现代码在文件 order.jsp 中。主要代码如下。

(1) 对用户是否登录进行检测部分代码如下：

```
<%String username="";
username=(String)session.getAttribute("consumer");
if(username==""||username==null){
out.println("<script language='javascript'>alert('请先登录');window.location.
href='index.jsp';</script>");
}else{Vector ct=(Vector)session.getAttribute("ct");
if(ct==null||ct.size()==0){
response.sendRedirect("cart-null.jsp");
}
}
%>
```

(2) 对用户填写的数据进行检测部分代码如下：

```
<script type="text/javascript">
 function check(){
if(document.order.Truename.value=="")
{
window.alert("真实姓名不为空!");
window.order.Truename.focus();
return false;
}
 if(document.order.address.value=="")
{
window.alert("地址不为空!");
window.order.address.focus();
return false;
}
order.submit();
```

```
        }
        </script>
</head>
<body>
```

(3) 设置用户动作和接收数据部分代码如下：

```
<form name="order" action="OrdServlet" method="post">
<%
    String username123="";
    username123= (String)session.getAttribute("consumer");
%>
    <input type="hidden" name="action" value="1"/>
    <input type="hidden" name="username123" value="<%=username123 %>"/>
    <table width="100%" height="339"  border="0" cellpadding="0" cellspacing="0">
    <tr>
    <td width="7%" height="26"> </td>
    <td height="26" colspan="2" class="word_deepgrey">注意：请您不要恶意或非法提
交订单以免造成不必要的麻烦！！！</td>
    </tr>
    <tr>
        <td height="26" colspan="2" align="center">用户名：</td>
        <td width="74%"><%=username123 %></td>
    </tr>
    <tr>
        <td height="26" colspan="2" align="center">真实姓名：</td>
        <td><input type="text" name="Truename" /> * </td>
</tr>
<tr>
<td height="26" colspan="2" align="center">联系地址：</td>
<td><input type="text" name="address" /> * </td>
</tr>
<tr>
<td height="101" colspan="2" align="center">备注：</td>
<td><textarea name="bz" cols="50" rows="5" class="textarea" id="bz">
</textarea></td>
</tr>
<tr align="center">
<td colspan="3"><input type="button" value="提交" onclick="check()"> 
<input name="Submit2" type="button" class="btn_grey" value="返回" onClick=
"history.back(1);"></td>
</tr>
</table>
</form>
</html>
```

单击用户功能菜单页面中的订单管理(即发出订单管理请求),进入用户订单管理页面,页面如图 6.56 所示。

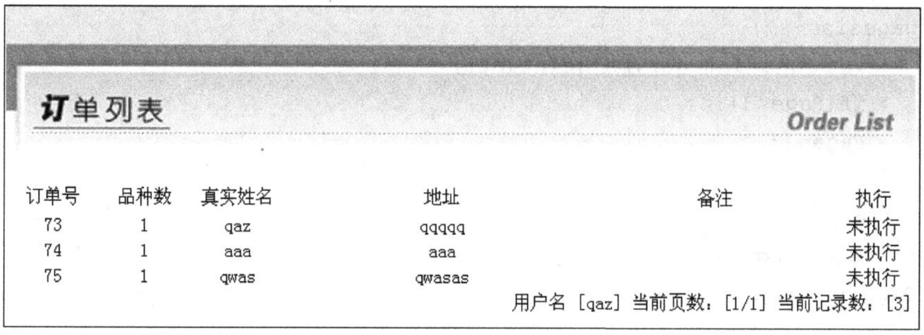

图 6.56 用户订单管理页面

用户订单管理页面实现代码在文件 order_user.jsp 中。主要代码如下。
(1) 检测用户是否登录部分代码如下:

```
<%
String username="";
    username=(String)session.getAttribute("consumer");
    if(username==""||username==null){
    out.println("<script language='javascript'>alert('请先登录!');window.
    location.href='index1.jsp';</script>");
    }else{
DBUtil conn=new DBUtil();
ResultSet rs = conn.executeQuery("select * from oorder where username = '" +
username+"'");
```

(2) 显示订单部分代码如下:

```
    String str=(String)request.getParameter("Page");
    if(str==null){
    str="0";
    }
    int pagesize=7;
        rs.last();
    int RecordCount=rs.getRow();
    if(RecordCount==0){
    %>
    </table>
<table width="100%" border="0" cellspacing="0" cellpadding="0">
    <tr>
    <td align="center">没有订单! 
    <%} %>
    <%
    if(RecordCount!=0){
```

```
            rs.first();
            int maxpage=0;
        maxpage=(RecordCount%pagesize==0)?(RecordCount/pagesize):(RecordCount/
        pagesize+1);
            int Page=Integer.parseInt(str);
            if(Page<1){
            Page=1;
            }
            else{
        if(Page>maxpage){
        Page=maxpage;
        }
        }
        rs.absolute((Page-1)*pagesize+1);                    //定位记录指针
         for(int i=1;i<=pagesize;i++){
        order ord=new order();
        ord.setorderID(rs.getInt("orderID"));
        ord.setbnumber(rs.getInt("bnumber"));
        ord.setTruename(rs.getString("Truename"));
        ord.setaddress(rs.getString("address"));
        ord.setbz(rs.getString("bz"));
        ord.setenforce(rs.getInt("enforce"));
         %>
        <tr>
        <td height="20" align="center"><%=ord.getorderID()%></td>
        <td align="center"><%=ord.getbnumber()%></td>
        <td align="center"><%=ord.getTruename()%></td>
        <td align="center"><%=ord.getaddress()%></td>
        <td align="center"><%=ord.getbz()%></td>
        <td align="center"><%if(ord.getenforce()==0){%>
        未执行
        <%}else{ %>已执行<%} %></td>
    </tr>
    <%
        try{
        if(!rs.next()){break;}
            }catch(Exception e){}
        }
        %>
    </table>
翻页部分代码略。
```

6.4.5 系统实现——BLL 层

在图 6.33 中 src 包含的 Servlet 包中的类提供了在页面上进行请求的响应和处理,

完成相对应的业务逻辑层功能,对应三层架构中的 BLL 层。因此这个包中定义的类是对相应 Web 页面上的请求进行响应处理并做出反馈,以实现设计阶段中设计的各种功能,包含的类如图 6.57 所示。

ConsServlet.Java 主要是针对与普通用户有关的页面上的请求进行响应并做出处理,FurServlet.Java 主要是针对与家具有关的页面上的请求进行响应处理并做出反馈,LoginServlet.Java 主要是针对与登录注册有关的页面上的请求进行响应并进行处理,OperServlet.Java 主要是对操作员的页面上的请求进行响应并进行处理;OrdServlet.Java 主要是对与订单相关的页面上的请求进行响应并进行处理;PlacardServlet.Java 主要是对针对公告相关页面上的请求进行响应并做相关处理。

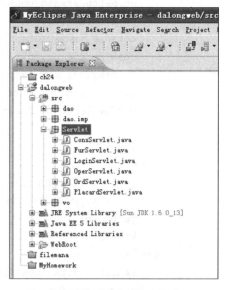

图 6.57　Servlet 包的构成

ConsServlet.Java 中与处理用户相关动作有关的部分代码(即对事件进行响应并做出处理)的主要代码如下:

```
switch(action){
    case 1:              //对用户功能界面中的修改个人信息请求的响应和服务
        request.getRequestDispatcher("consumermana/changecons.jsp").forward(request,response);
        break;
    case 2:              //修改个人信息界面的修改请求的响应和服务
        String id=request.getParameter("id");
        String cname=request.getParameter("cname");
        String cpassword=request.getParameter("password");
        String ctel=request.getParameter("ctel");
        userdao udao=DAOFactory.instance().getuserdao();
        consumer uu=new consumer();
        uu=udao.getId(cname,1);
        if(uu!=null){
            request.setAttribute("iderror","此用户名已存在!");
            try{
                //转发至注册失败页面
                request.getRequestDispatcher("consumermana/changecons.jsp").forward(request,response);
            }catch(Throwable t){
                getServletContext().log(t.getMessage());
            }
        }
        else{
```

```
            uu.setcname(cname);
            uu.setcpassword(cpassword);
            uu.setctel(ctel);
            consdao cdao=DAOFactory.instance().getconsdao();
            cdao.update(uu,id);
            PrintWriter out=response.getWriter();
            out.println("<script language='javascript'>alert('信息修改成功');
            window.location.href='index.jsp';</script>");
                }
            break;
        case 3:                              //对用户功能界面中的注销请求的响应和服务
            session.invalidate();    //取消所有的回话
            request.getRequestDispatcher("homepage.jsp").forward(request,
            response);
            break;
        case 4:                              //对用户功能界面中的购物车请求的响应和服务
            request.getRequestDispatcher("cart/cart-see.jsp").forward
            (request,response);
            break;
        case 5:                              //对用户功能界面中的订单查询请求的响应和服务
            request.getRequestDispatcher("ordermana/order_user.jsp").forward
            (request,response);
            break;
        }
```

FurServlet.Java中与处理用户动作有关的部分代码(即对事件进行响应并做出处理)的具体代码如下:

```
        switch(action){
        case 1:                              //对商品管理页面中的修改请求的响应和服务
            String fidf=request.getParameter("id");
            int fid=Integer.parseInt(fidf);
            furdao fdao=DAOFactory.instance().getfurdao();
            furniture furchange=fdao.getId(fid);
            request.setAttribute("furchange",furchange);

            request.getRequestDispatcher("furmana/furchange.jsp").forward
            (request,response);
            break;
        case 2:                              //对商品管理页面中修改页面的修改请求的响应和服务
            String fids=request.getParameter("fid");
            int fida=Integer.parseInt(fids);
            String fname=request.getParameter("fname");
            String fsummary=request.getParameter("fsummary");
            String fnumf=request.getParameter("fnum");
```

```java
            int fnum=Integer.parseInt(fnumf);
            String fmoneyf=request.getParameter("fmoney");
            int fmoney=Integer.parseInt(fmoneyf);
            furniture f=new furniture();
            f.setfid(fida);
            f.setfname(fname);
            f.setfsummary(fsummary);
            f.setfnum(fnum);
            f.setfmoney(fmoney);
            fdao=DAOFactory.instance().getfurdao();
            fdao.update(f);
            PrintWriter out=response.getWriter();
            out.println("修改成功");
            break;
        case 3:                          //对商品管理页面中的删除请求的响应和服务
            fids=request.getParameter("id");
            fdao=DAOFactory.instance().getfurdao();
            out=response.getWriter();
            if(fdao.delete(fids)){
                out.print("<script language='javascript'>alert('删除成功!');
                </script>");
                request.getRequestDispatcher("furmana/furmana.jsp").forward
                (request,response);
                }
            else
                {
                out.print("<script language='javascript'>alert('删除失败!');
                </script>");
                request.getRequestDispatcher("furmana/furmana.jsp").forward
                (request,response);
                }
            break;
        case 4:                          //对商品管理页面中的添加请求的响应和服务

            //out=response.getWriter();
            //out.println("dalong");
            fname=request.getParameter("fname");
            fsummary=request.getParameter("fsummary");
            String fmoney1=request.getParameter("fmoney");
            fmoney=Integer.parseInt(fmoney1);
            String fpic=request.getParameter("fpic");
            fpic="./images/"+fpic+".jpg";

            furniture fur=new furniture();
```

```
                fur.setfname(fname);
                fur.setfsummary(fsummary);
                fur.setfmoney(fmoney);
                fur.setfpic(fpic);
                fdao=DAOFactory.instance().getfurdao();
                out=response.getWriter();
                if(fdao.insert(fur)) out.print("<script language='javascript'>
                alert('添加成功!');window.location.href='furmana/furmana.jsp';
</script>");
                else out.print("<script language='javascript'>alert('添加失败!');
                window.location.href='furmana/furmana.jsp';</script>");
                break;
            case 5:                             //对商品管理页面中的翻页请求的响应和服务
                fname=request.getParameter("fname");
                request.setAttribute("fname", fname);
                furniture fu=new furniture();
                fu.setfname(fname);
                int currentPage=1;
                try{
                    currentPage=Integer.parseInt(request.getParameter("page"));
                }catch(NumberFormatException e){
                    currentPage=1;
                }
                request.setAttribute("currentPage", currentPage);
                int pageMaxNum=5;
                request.setAttribute("pageMaxNum", pageMaxNum);
                furdao fudao=DAOFactory.instance().getfurdao();
                int count=fudao.getCount(fu);
                int totalPage = (count%pageMaxNum==0)? count/pageMaxNum: (count/
                pageMaxNum+1);
                request.setAttribute("totalPage", totalPage);
                List<furniture> flist= fudao.getByCondition(fu,(currentPage-1)*
                pageMaxNum, pageMaxNum);
                request.setAttribute("flist", flist);
                request.getRequestDispatcher("main.jsp").forward(request,
                response);
                break;
            case 6:                             //对购买请求的响应和服务,分没用用户登录和有用户登录
                String username="";
                username=(String)session.getAttribute("consumer");
                if(username==""||username==null){
                out=response.getWriter();
                out.println("<script language='javascript'>alert('请先登录!');
                window.location.href='index.jsp';   </script>");
```

```
        }else{
            String ID1=request.getParameter("id");
            int ID=Integer.parseInt(ID1);
            fudao=DAOFactory.instance().getfurdao();
            furniture fur1=new furniture();
            fur1=fudao.getId(ID);
            int price=fur1.getfmoney();
            int num=1;
            cart ca=new cart();
            ca.setID(ID);
            ca.setprice(price);
            ca.setnum(num);
            boolean Flag=true;
//          HttpSession session=request.getSession();
            Vector ct=(Vector)session.getAttribute("ct");
            if(ct==null)
            {
                ct=new Vector();
            }else{
                for(int j=0,i=0;i<ct.size();i++)
                {
                    cart cb=(cart)ct.elementAt(i);
                    if(cb.getID()==ca.getID()){
                        j=cb.getnum();
                        j++;
                        cb.setnum(j);
                        ct.setElementAt(cb,i);
                        Flag=false;
                    }
                }
            }
            if(Flag)ct.addElement(ca);
            session.setAttribute("ct", ct);
            response.sendRedirect("cart/cart-see.jsp");
        }
        break;
    }
```

LoginServlet.Java 中与处理用户动作有关的部分代码(即对事件进行响应并做出处理)的主要代码如下:

```
    switch(action){
    case 1:                         //对登录界面中的登录请求的响应和服务
        String name=request.getParameter("name");
        String password=request.getParameter("password");
```

```java
                String r=request.getParameter("role");
                int role=Integer.parseInt(r);
                boolean b=udao.isLogin(name, password,role);
                if(!b){
                        request.setAttribute("error","用户名或密码错误!");
                        request.getRequestDispatcher("login.jsp").forward(request,
                        response);
                }
                else {
                    if(role==1){
                    u=udao.getId(name,role);
                    HttpSession session=request.getSession();
                    session.setAttribute("consumer", name);
request.getRequestDispatcher("consumermana/cons.jsp").forward(request,
response);
                }
                    else
                    {
                        v=udao.getId1(name,role);
                        HttpSession session=request.getSession();
                        session.setAttribute("operator", name);
    request.getRequestDispatcher("operatormana/oper.jsp").forward(request,
    response);
                    }
                }
    break;
    case 2:                              //对注册页面中的返回请求的响应和服务
        request.getRequestDispatcher("login.jsp").forward(request, response);
        break;
    case 3:                              //对登录界面中的注册请求的响应和服务
        request.getRequestDispatcher("register.jsp").forward(request, response);
        break;
    case 4:                              //对注册页面中的立即注册请求的响应和服务
        String cname=request.getParameter("cname");
        String pwd=request.getParameter("password");
        String tel=request.getParameter("tel");
        role=1;
        try{
            u=new consumer();
            u=udao.getId(cname,role);
            if(u!=null){
                request.setAttribute("iderror","此用户名已存在!");
                try{
```

```
            //转发至注册失败页面
                   request.getRequestDispatcher("register.jsp").forward(request,
                   response);
                }catch(Throwable t){
                   getServletContext().log(t.getMessage());
                }
         }else{
            u=new consumer();
            u.setcname(cname);
            u.setcpassword(pwd);
            u.setctel(tel);
            udao.insert(u);
            request.setAttribute("consumer", u);
            try{
                //转发至注册成功页面
                   request.getSession(true).setAttribute("name", cname);
                   request. getRequestDispatcher ( " index. jsp "). forward ( request,
                   response);
                }catch(Throwable t){
                   getServletContext().log(t.getMessage());
                }
         }
      }catch(Exception e){
         e.printStackTrace();
      }
      break;
}
```

OperServlet.Java 中与处理用户动作有关的部分代码(即对事件进行响应并做出处理)的具体代码如下：

```
      switch(action){
      case 1:                          //对操作员功能界面中的家具管理请求的响应和服务
            request. getRequestDispatcher ( " furmana/furmana. jsp "). forward
            (request, response);
            break;
      case 2:
            break;
      case 3:                          //对操作员功能界面中的订单管理请求的响应和服务
            request.getRequestDispatcher("ordermana/ordermanage.jsp").forward
            (request, response);
            break;
      case 4:                          //对操作员功能界面中的注销请求的响应和服务
```

```
        session.invalidate();
        request.getRequestDispatcher("homepage.jsp").forward(request,
        response);
        break;
    case 5:                          //对操作员功能界面中的公告管理请求的响应和服务

        request.getRequestDispatcher("placardmana/placard_mana.jsp").
        forward(request,response);
        break;
    case 6:
//对操作员功能界面中的会员管理请求的响应和服务,包括会员管理界面中的删除请求
        String cname=request.getParameter("cname");
        userdao udao=DAOFactory.instance().getuserdao();
        PrintWriter out=response.getWriter();
        if(udao.delete(cname))out.print("<script language='javascript'>
        alert('删除成功!');window.location.href='operatormana/membermana.
        jsp';</script>");
        else out.print("<script language='javascript'>alert('删除失败!');
        window.location.href='operatormana/membermana.jsp';</script>");
        break;
    case 7:                          //按家具名查询家具请求的响应和服务
        String d="fname";
        request.setAttribute("k",d);

          request.getRequestDispatcher("index.jsp").forward(request,
          response);
        break;
    case 8:                          //按家具描述查询家具请求的响应和服务
        d="fsummary";
        request.setAttribute("k",d);

        request.getRequestDispatcher("index.jsp").forward(request,
        response);
        break;
    case 9:                          //按家具编号查询家具请求的响应和服务
         d="fnum";
        request.setAttribute("k",d);

        request.getRequestDispatcher("index.jsp").forward(request,
        response);
        break;
    case 10:                         //按家具价格查询家具请求的响应和服务
        d="fmoney";
        request.setAttribute("k",d);
```

```
        request. getRequestDispatcher ( " index. jsp "). forward ( request,
        response);
        break;
    case 11:                       //无条件查询家具请求的响应和服务
        d="";
        request.setAttribute("k",d);

        request. getRequestDispatcher ( " index. jsp "). forward ( request,
        response);
        break;
}
```

OrdServlet.Java中与处理用户动作有关的部分代码(即对事件进行响应并做出处理)具体代码如下:

```
switch(action){
    case 1:                         //生成订单,即对购物车界面中的提交请求的响应和服务
        String username=request.getParameter("username123");
        String Truename=request.getParameter("Truename");
        String address=request.getParameter("address");
        String bz=request.getParameter("bz");
        int orderID=0;
        int number=0;
        int price1=0;
        int sum=0;
        int Totalsum=0;
        String flag="True";
        int temp=0,temp1=0;
        int ID2=-1;
        HttpSession session=request.getSession();
        Vector ct1=(Vector)request.getSession().getAttribute("ct");

        for(int i=0;i<ct1.size();i++)
        {
            cart cc=(cart)ct1.elementAt(i);
            price1=cc.getprice();
            number=cc.getnum();
            sum=number * price1;
            Totalsum+=sum;
        }
        consumer cons=new consumer();
        consdao cdao=DAOFactory.instance().getconsdao();
        int cmoney=0;
        PrintWriter out=response.getWriter();
```

```
cmoney=cdao.getmoney(username);
if(cmoney<Totalsum)
{
    flag="false";
}else{
    cmoney=cmoney-Totalsum;
    cdao.updatem(username, cmoney);
}
if(flag.equals("false")){
    out=response.getWriter();
    out.println("<script language='javascript'>alert('订单无效');
    history.back();</script>");
}else{
int bnumber=ct1.size();
order ord=new order();
ord.setbnumber(bnumber);
ord.setusername(username);
ord.setTruename(Truename);
ord.setaddress(address);
ord.setbz(bz);
orderdao odao=DAOFactory.instance().getorderdao();
odao.insert(ord);
orderdao odao1=DAOFactory.instance().getorderdao();
furdao fdao=DAOFactory.instance().getfurdao();
temp=odao1.getId(ord);
if(temp==0){
    flag="false";
}else{
    orderID=temp;
}
order_detail ordl=new order_detail();
String str="";
for(int i=0;i<ct1.size();i++)
{
    cart cc=(cart)ct1.elementAt(i);
    ID2=cc.getID();
    price1=cc.getprice();
    number=cc.getnum();
    sum=number * price1;
    ordl.setorderID(orderID);
    ordl.setgoodsID(ID2);
    ordl.setprice(price1);
    ordl.setnumber(number);
    temp=odao.insert1(ordl);
```

```
                temp1=fdao.update(ID2, number);
                Totalsum+=sum;
                if(temp==0||temp==0)
                {
                    flag="false";
                }
            }
            if(flag.equals("false")){
                out=response.getWriter();
                out.println("<script language='javascript'>alert('订单无效');
                history.back();</script>");
            }else{
                session.removeAttribute("ct");
                out=response.getWriter();
                out.println("<script language='javascript'>alert('订单生成');
                window.location.href='index.jsp'  </script>");
            }
        }
        break;
    case 2:               //删除订单,即操作员界面中订单管理中的删除请求的响应和服务
        String sdate=request.getParameter("sdate");
        String edate=request.getParameter("edate");
        count ct=new count();
        int currentPage=1;
        try{
            currentPage=Integer.parseInt(request.getParameter("page"));
        }catch(NumberFormatException e){
            currentPage=1;
        }
        request.setAttribute("currentPage", currentPage);
        int pageMaxNum=10;
        request.setAttribute("pageMaxNum", pageMaxNum);
        orderdao oodao=DAOFactory.instance().getorderdao();
        int count1=oodao.getCount(ct,sdate,edate);
        int totalPage= (count1%pageMaxNum==0)? count1/pageMaxNum: (count1/
        pageMaxNum+1);
        request.setAttribute("totalPage", totalPage);
        List< count > clist = oodao.getByCondition (ct, (currentPage - 1) *
        pageMaxNum, pageMaxNum, sdate, edate);
        request.setAttribute("clist", clist);
        int ssum=oodao.getSum(sdate, edate);
        request.setAttribute("ssum", ssum);

        request.getRequestDispatcher ( " furmana/furallinfo. jsp"). forward
```

```
            (request,response);
            break;
        }
```

PlacardServlet.Java 中与处理用户动作有关的部分代码(即对事件进行响应并做出处理)的具体代码如下：

```
        switch(action){
        case 1:                    //发布公告,即公告管理界面中的添加请求的响应和服务
            placard pla=new placard();
            pla.settitle(request.getParameter("title"));
            pla.setcontent(request.getParameter("content"));
            placarddao plad=DAOFactory.instance().getplacarddao();
            plad.insert(pla);
            PrintWriter out1=response.getWriter();
            out1.println("<script language='javascript'>alert('发布成功!');
</script>");

            request.getRequestDispatcher("placardmana/placard_mana.jsp").
            forward(request, response);
            break;
        case 2:    //显示公告详细内容,即对公告界面中单击某一公告题目请求的响应和服务
            int ID=-1;
            placard pla1=new placard();
            ID=Integer.parseInt(request.getParameter("ID"));
            placarddao plad1=DAOFactory.instance().getplacarddao();
            if(ID>0){
                pla1=plad1.query(ID);
                request.setAttribute("pla1", pla1);
            }

            request.getRequestDispatcher("placardmana/placard_detail.jsp").
            forward(request, response);
            break;
        case 3:                    //即对公告管理界面中删除请求的响应和服务
            String ID1[]=request.getParameterValues("delid");
            String bbsID="";
            PrintWriter out=response.getWriter();
            if(ID1.length>0){
                for(int i=0;i<ID1.length;i++){
                    bbsID=bbsID+ID1[i]+",";
                }
                bbsID=bbsID.substring(0,bbsID.length()-1);
                                    //去除连接字符串中最后一个字符即逗号
                int ret=-1;
```

```
            placarddao plad2=DAOFactory.instance().getplacarddao();
            ret=plad2.delete(bbsID);
            if(ret==0){
                out.println("<script language='javascript'>alert('公告信息
                删除失败!');</script>");

                request.getRequestDispatcher("placardmana/placard_mana.
                jsp").forward(request,response);
            }
            else{
                out.println("<script language='javascript'>alert('公告信息
                删除成功!');</script>");

                request.getRequestDispatcher("placardmana/placard_mana.
                jsp").forward(request,response);
            }
        }
     else{
        out.println("<script language='javascript'>alert('操作有误!');
        </script>");

        request.getRequestDispatcher("placardmana/placard_mana.jsp").
        forward(request,response);
     }
     break;
}
```

第 7 章 网络办公自动化系统的研究与实现

办公自动化简称 OA,是办公信息处理流程的自动化。办公自动化系统利用计算机技术,使人与人面对面的各种办公活动逐步向人与各种机器之间的办公活动转化,从而可以充分利用各种信息,使工作效率和工作质量得以极大提高,从而达到提高生产率的目的。

办公自动化的发展得益于计算机技术的发展,尤其是网络技术与数据库技术的发展,使人们的生活与工作方式发生了很大的改变。网络技术的发展使计算机之间的通信、信息的共享成为可能,而数据库技术的应用则为人们提供了数据存储、信息查询、信息编辑等功能,从而使工作更加高效地进行。数据库技术始于 20 世纪 60 年代,历经 40 多年的发展,现在已经形成了完善的理论体系,并成为计算机软件方面的一个重要分支。数据库技术体现了现代先进的数据管理方法和方式,使计算机的应用真正渗透到社会各个部门,在数据相关处理领域发挥着越来越大的作用。互联网技术的出现,进一步丰富了人类的生活,使数字化技术一步步走进人们的生活与工作。互联网技术与数据库技术的相互结合使办公自动化系统在人类生活中得到更为广泛的应用。

本章介绍的办公自动化网站的设计与实现采用 MVC 三层架构,软件开发环境如下所示。

(1) 操作系统:Windows XP。
(2) 开发环境:Microsoft Visual Studio 2010。
(3) 数据库:SQL Server 2008。

7.1 需求调查分析

本章介绍的办公自动化系统不是一个实际的建设项目,系统目标和功能是研究人员参照一般办公自动化系统、结合教学过程中的实际情况而设定的。采用面向对象的开发方法,主要解决办公自动化中涉及个人信息管理、文件的收发、公共的收发等核心问题。

7.1.1 系统定义及可行性分析

网络办公自动化系统是针对小型企事业单位设计的一款通用办公系统,其主要用户包括管理员、部门领导和普通员工三类,主要完成各类信息的管理、个人计划的制定、文件的收发及公告的发布及查看。管理员可以完成个人信息、员工信息、部门信息、活动信息和公共信息的管理,部门领导可完成个人信息、员工信息、活动信息和公共信息的管理,而普通员工只能完成个人信息的管理、活动信息和公共信息的管理。

1. 技术可行性

随着网络技术和数据库技术支持都日趋成熟,作为服务器具备的硬件条件是绝大多数的商业计算机都具备的。ASP.NET 网络编程语言实现图形化、简易化的表现和管理,C♯语言作为后台运行处理,利用 ASP.NET 技术与 SQL Server 2008 数据库的结合是网络系统设计的常用模式,所以技术成熟,安全性强。因此,从技术角度上考虑,本系统是可行的。

2. 经济可行性

以前公文和文件的流转、信息传递与录入全部依靠人工操作,这些传统的办公流程极大地浪费了企业的时间和精力,而且办公处理很不及时,导致管理流程比较复杂和低效。所以,无论是企业领导者还是员工都迫切需要高效、简捷、方便的办公方式,因此办公自动化系统应运而生。办公自动化系统主要成本集中在系统软件的开发上,但是当系统投入使用后可以使企业节约大量的人力、财力、物力,为企业所带来的效益远远超出系统软件的开发成本。因此,从经济角度上考虑,本系统是可行的。具体开发费用和使用收益如表 7.1 所示。

表 7.1 办公自动化系统费用和收益表

开 发 费 用	运 行 费 用
设计人员工资	软件许可费
开发培训	软硬件维护
购买硬件和软件	使用人员工资
占用办公场地和设备	沟通费用
有 形 收 益	无 形 收 益
办公使用实体费用降低	工作人员工作量的减少
工作人员的工时降低	对于社会的帮助
场地费用的降低	管理流程规范化

3. 操作可行性

系统设计时充分考虑了用户的需求,人机交互界面友好,操作流程简单;数据处理迅速、准确、可靠;可用性强;容易扩充。所以对于用户的使用水平要求并不高,因此,从操作角度上考虑,本系统是可行的。

7.1.2 系统需求分析和目标设定

对现有办公自动化系统进行分析和研究,得出对系统需求如下:
(1) 建立一个相对完善的网上办公系统;
(2) 在网站中相关信息表达准确、显示方式恰当、布局合理;
(3) 记载的信息准确,方便使用;保证信息的安全;
(4) 功能全面;
(5) 方便一般用户完成相关操作,各种操作合乎用户的一般习惯;
(6) 方便管理人员完成相关操作,各种操作合乎管理员的操作习惯;
(7) 各种操作流程合理;
(8) 提高办公效率。
对上述需求进行进一步分析、整理,提出了具体的系统目标为:
(1) 梳理出先进、规范的业务流程;
(2) 快速、准确的信息收集,能真正实现决策支持;
(3) 通过对网站的管理,能逐步提高办公效率;
(4) 加强系统流程控制,保证系统和数据安全;
(5) 实现数据的集中化、数字化处理。
为了到达上述目标,系统要实现的主要功能如下。

1. 功能需求

1) 个人管理
(1) 管理任务计划:可以添加、删除任务以及编辑任务状态。
(2) 个人信息管理:可以查看、编辑个人身份信息。
(3) 密码管理:可以修改个人系统密码等。

2) 部门管理(只有高级用户可操作)
(1) 查看部门信息:可以查看各个部门信息。
(2) 编辑职位信息:可以删除部门、删除职位、添加职位。
(3) 新建部门:可以添加新部门。

3) 员工管理(高级用户和中级用户可操作)
(1) 查找员工信息:可以查看员工信息。
(2) 编辑员工信息:可以修改员工信息。
(3) 添加员工信息:可以添加新员工。

4) 公告管理
(1) 查看公告:可以查看公告、删除公告。
(2) 发布公告:可以发布新公告。

5) 活动管理
(1) 查看活动:可查看活动信息。
(2) 添加活动:可以添加新活动。
(3) 活动投票:可以进行活动投票。
(4) 发送文件:可以发送文件。

(5) 接收文件：可以接收文件、删除文件。

2．非功能需求

(1) 界面需求：用户界面友好、简洁，操作简单，方便。

(2) 性能需求：信息的存取与检索要快捷，稳定性要强，安全性要高。

(3) 权限需求：在用户进行登录时，只有授权的账号和密码才能进入。

7.2 用例建模

7.2.1 角色用例图

在以上分析的基础上，可以把系统分为普通用户、中级用户、高级用户这三大角色。

高级用户角色：拥有系统的全部权限，具体用例图如图7.1所示。

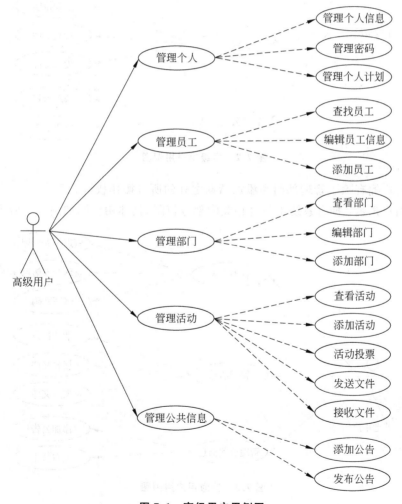

图7.1　高级用户用例图

高级用户可以使用系统的全部功能,这些功能包括管理个人信息、管理员工、管理部门、管理活动和管理公共信息,其中每个功能下又包含许多子功能。

中级用户角色:拥有该系统部分权限及其一般权限,具体用例图如图7.2所示。

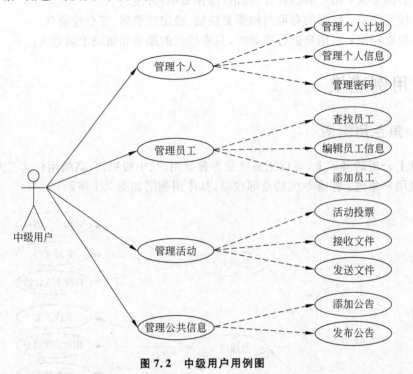

图7.2 中级用户用例图

中级用户拥有除了管理部门和添加活动之外的所有操作权限。

普通用户角色:拥有系统中与之相关的部分权限,具体用例图如图7.3所示。

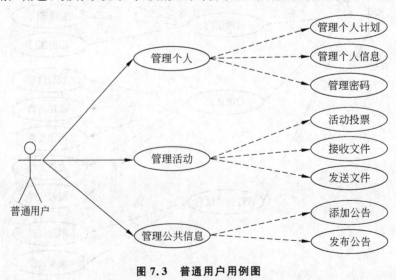

图7.3 普通用户用例图

普通用户拥有的权限最少,普通用户不具有管理部门和管理员工及添加活动权限,但该角色可以进行除此之外的所有操作。

7.2.2 模块用例图

根据以上分析,可以把系统分为管理个人计划、管理个人信息、管理密码、活动投票、添加活动、查看活动、查找员工、添加员工、编辑员工信息、查看部门、添加部门、编辑部门信息、发送文件、接收文件、发布公告、查看公告人事管理系统的登录模块,具体用例图如图 7.4～图 7.20 图。

1. 管理个人计划

用户可以管理个人计划,用户在管理个人计划模块中可以添加个人任务,并可编辑任务状态,用户觉得任务无须提醒时可删除任务。

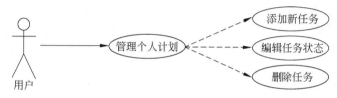

图 7.4　管理个人计划用例图

2. 管理个人信息

用户在管理个人信息时,可以查看个人信息并填写个人信息。

图 7.5　管理个人信息用例图

3. 管理密码

用户可以修改自己的登录密码,修改密码操作需经过验证旧密码和验证新密码两个过程。

图 7.6　管理密码用例图

4. 活动投票

用户可进行活动投票,系统将显示活动信息,以便用户进行投票。

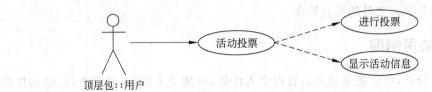

图 7.7　活动投票用例图

5．添加活动

高级用户可以在活动信息后添加活动。

图 7.8　添加活动用例图

6．查看活动

用户可以查看活动，并可在显示的活动信息后选择删除活动。

图 7.9　查看活动用例图

7．查找员工

中级和高级用户可以通过员工名称来查找员工信息。

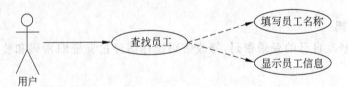

图 7.10　查找员工用例图

8．添加员工

中级和高级用户可以在填写员工信息后添加新员工。

图 7.11　添加员工用例图

9. 编辑员工信息

中级和高级员工可以变更员工信息,并可删除该员工。

图 7.12 编辑员工用例图

10. 查看部门

高级用户可以查看企业各部门的信息,用户选择部门后即可显示该部门信息。

图 7.13 查看部门用例图

11. 添加部门

高级用户可以在填写新部门信息后添加新部门,也可选择某一部门修改其信息。

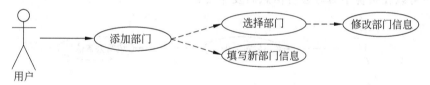

图 7.14 添加部门用例图

12. 编辑职位信息

高级用户可以编辑职位信息,包括修改职位信息、添加新职位信息、删除部门和删除职位。

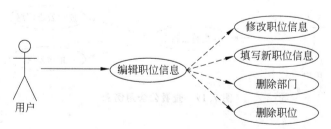

图 7.15 编辑职位信息用例图

13. 发送文件

用户可以填写文件相关信息后选择目标,向其发送文件。

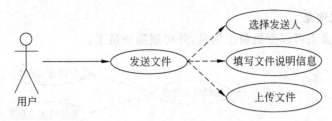

图 7.16 发送文件用例图

14. 接收文件

用户可以接收别人发送给其的文件,并可查看文件信息并删除文件。

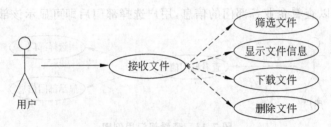

图 7.17 接收文件用例图

15. 发布公告

用户可以在系统中填写公告信息后发布公告。

图 7.18 发布公告用例图

16. 查看公告

用户可以查看公告和删除公告。

图 7.19 查看公告用例图

17. 登录

用户填写用户信息和验证码信息经过验证后可登录系统。

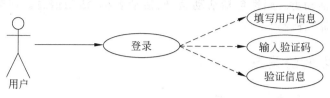

图 7.20　登录模块用例图

7.3　静态建模

7.3.1　系统类图

根据对系统功能和用例模型的分析，提取出系统中主要包含的类有员工、部门、公告、计划、活动等，分析这些类的主要属性、操作和类之间的关系，得出系统基础类图如图 7.21 所示。

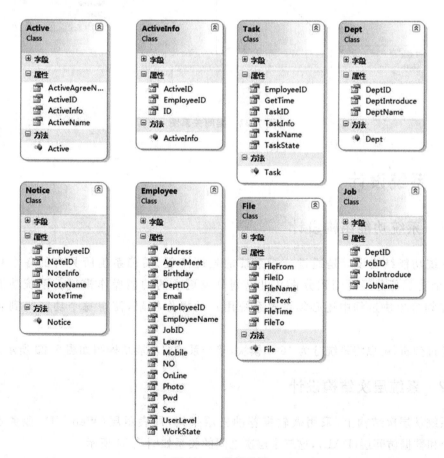

图 7.21　基础类图

Active 类是活动类。属性包括活动编号、活动名称、活动信息。该类的功能主要是对活动信息的存储、查询和更新。

7.3.2 各类之间的关系

各类之间的关系如图 7.22 所示。

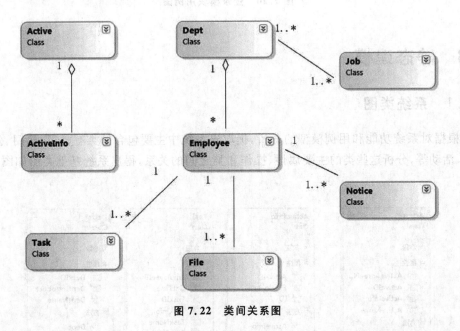

图 7.22 类间关系图

7.4 系统设计

7.4.1 系统功能结构设计

系统功能结构设计是通过分析系统的处理流程,明确系统所具有的各个功能部分,每个功能部分经过细致分析又可以划分成更小的功能操作模块。系统功能结构设计在划分模块过程中还必须设置系统用户的操作权限,使得每个功能得到合理的使用。

经过分析,可以将系统分为 16 个模块,整个系统的功能结构图如图 7.23 所示。

7.4.2 系统层次结构设计

系统从层次结构上,采用微软推荐的三层架构,即表示层(Web UI)、业务逻辑层(BLL)和数据访问层(DAL),这三个层次之间的关系如图 7.24 所示。

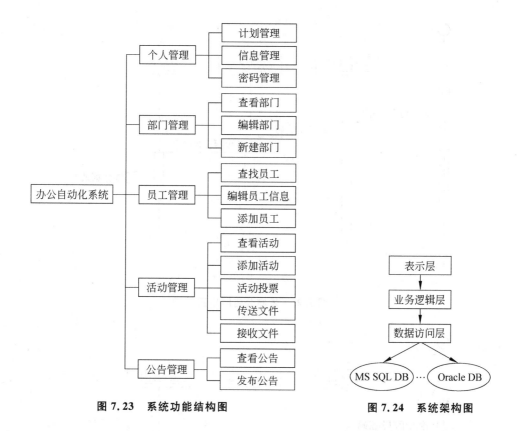

图 7.23 系统功能结构图　　　　图 7.24 系统架构图

7.5 动态建模

动态建模是建立动态模型,动态模型明确规定什么时候做,主要由状态图、活动图、顺序图和协作图描述。

7.5.1 模块时序图

由于篇幅的限制,下面重点介绍发布公告、发送文件、添加部门、管理密码的时序图,具体时序图如图 7.25～图 7.28 所示。

发布公告:普通用户、中级用户、高级用户三者都具有发布公告权限。用户在输入用户名和密码经过验证成功后进入主管理界面,在发布公告界面填写公告信息可发布公告。具体时序图过程如图 7.25 所示。

发送文件:用户可以发送文件给指定对象,用户经过登录验证后进入主管理界面,填写文件相关信息可发送文件,并将文件信息录入数据库。具体时序图过程如图 7.26 所示。

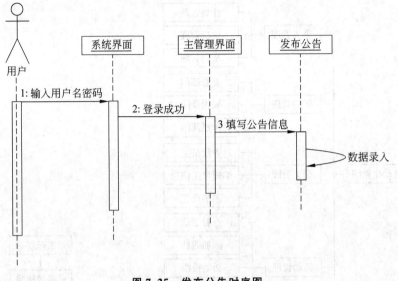

图 7.25 发布公告时序图

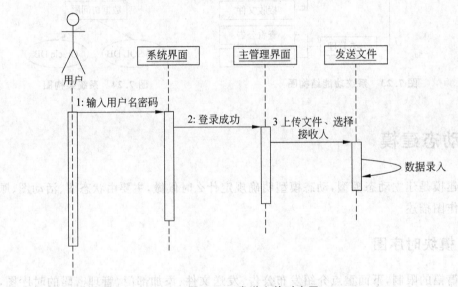

图 7.26 发送文件时序图

添加部门:高级用户具有添加部门权限,高级用户经过登录验证后进入主管理界面,填写部门信息可添加部门。具体时序图过程如图 7.27 所示。

管理密码:三种用户都具有密码管理的权限,用户经登录验证后进入主管理界面,选择管理密码可修改密码。具体时序图过程如图 7.28 所示。

7.5.2 模块活动图

由于篇幅的限制,下面重点介绍接收文件、编辑职位信息、编辑员工信息、管理个人计划的活动图,具体活动图如图 7.29~图 7.32 所示。

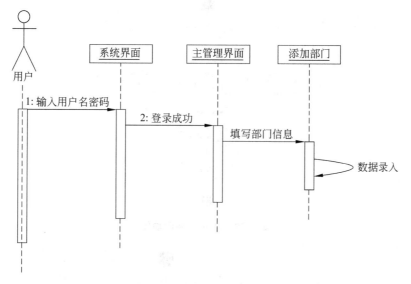

图 7.27　添加部门时序图

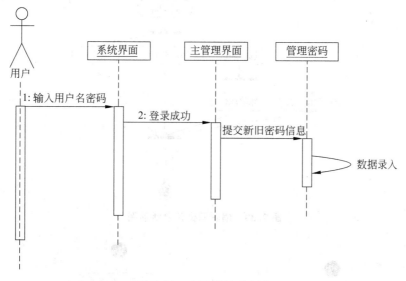

图 7.28　密码管理时序图

接收文件：普通用户、中级用户、高级用户拥有接收文件权限，用户在接收文件模块可查看文件信息、下载文件和删除文件，确认后完成操作。具体活动图过程如图 7.29 所示。

编辑职位信息：只有高级用户可以编辑职位信息，此模块高级用户在选择某一部门后可删除部门、添加新职位、修改职位信息和删除职位。具体活动图过程如图 7.30 所示。

编辑员工信息：高级用户和中级用户拥有编辑员工信息权限，用户选中员工后可修改该员工信息和删除该员工，确认后完成操作。具体活动图过程如图 7.31 所示。

管理个人计划：高级用户、中级用户、普通用户三者拥有管理个人计划权限，用户可添加新任务，编辑任务状态和删除任务。具体活动图过程如图 7.32 所示。

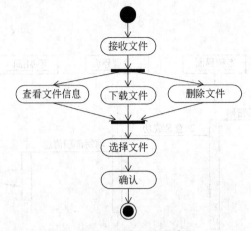

图 7.29　接收文件活动图

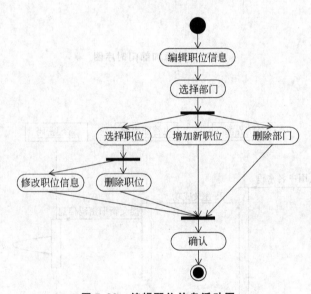

图 7.30　编辑职位信息活动图

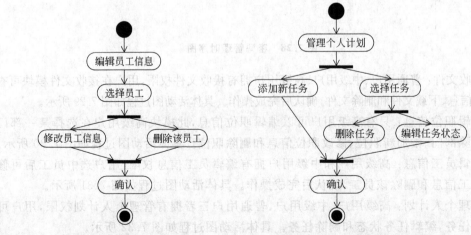

图 7.31　编辑员工活动图　　　　图 7.32　个人计划管理活动图

7.6 输入输出设计

输入输出设计是办公自动化系统与用户交互的界面。它能够为用户建立良好的工作环境,激发用户努力工作、主动工作的热情;它符合用户习惯,方便用户操作,使办公自动化系统易于为用户所接受;它为用户提供易读易懂的信息形态。所以输入输出设计对用户来说,显得尤为重要。

7.6.1 输入设计

输入设计是办公自动化系统与用户之间交互的纽带,设计的任务是根据具体业务要求,确定适当的输入形式,使办公自动化系统获取工作中产生的正确信息。输入设计的目的是提高输入效率,减少输入错误。

(1) 输入方式的选择:输入方式是以键盘、鼠标为媒介将数据输入。系统管理员登录系统后可以维护基础数据,通过键盘、鼠标将数据录入系统中,确认保存将数据存入数据库相应的数据表中。

(2) 输入界面选择:输入界面的设计十分重要,为了增加用户的体验度、简单方便用户操作以及增强交互界面的美观性,因此本系统采用 ASP.NET 和 AJAX 控件建立用户交互界面。

7.6.2 输出设计

输出设计的任务是使办公自动化系统输出满足用户需求的信息,是系统实施的结果和目的。信息能够满足用户需求,关系到系统使用效果和系统的成功与否。以为用户提供及时、准确、全面的信息服务、便于阅读和理解符合用户习惯为原则,进行该系统的输出设计。

7.7 物理建模

7.7.1 系统部署

物理建模是建立物理模型,物理模型表述系统物理了点的分布,主要用部署图和构件图描述。该系统采用 B/S 架构,服务器端使用 IIS 部署网站程序,使用 SQL Server 2008 管理数据库。

7.7.2 数据库设计

数据库关系图描述了各表字段之间主外键的关系,数据库中有八个表,表 EMPLOYEES 通过属性 employeeID 和表 FILE、NOTICE、DEPT、ACTIVEINFO 联系,

ACTIVE 表和 ACTIVEINFO 表通过属性 activeID 联系,JOB 表和 DEPT 表通过属性 deptID 联系,具体关系图如图 7.33 所示。

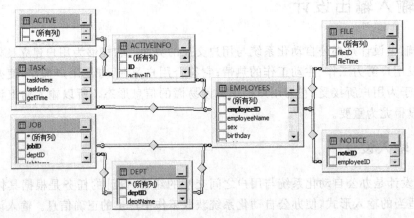

图 7.33　DatabaseOA 数据库关系图

7.7.3　数据库表设计

表设计关系到数据库的建立,根据以上逻辑设计的数据库关系图,具体表设计如表 7.2~表 7.9 所示。

表 7.2　ACTIVE 表的结构

字段名称	数据类型	空/非空	是否主键	约束条件	字段描述
activeID	int	非空	是	自增 1	活动 ID
activeName	Varchar(50)	非空	否		活动名称
activeInfo	Varchar(10)	空	否		活动介绍
activeAgreeNum	int	非空	否		活动票数

表 7.3　ACTIVEINFO 表的结构

字段名称	数据类型	空/非空	是否主键	约束条件	字段描述
ID	int	非空	是	自增 1	ID
activeID	int	非空	否		活动 ID
employeeID	int	非空	否		用户 ID

表 7.4　DEPT 表的结构

字段名称	数据类型	空/非空	是否主键	约束条件	字段描述
deptID	int	非空	是		部门 ID
deptName	Varchar(50)	非空	否		部门名称
deptText	Varchar(50)	空	否		部门介绍

表 7.5 FILE 表的结构

字段名称	数据类型	空/非空	是否主键	约束条件	字段描述
fileID	int	非空	是	自增 1	文件 ID
fileTime	datetime	空	否		文件发送时间
fileText	Varchar(50)	空	否		文件描述
fileFrom	int	空	否		文件发送人 ID
fileTo	int	空	否		文件接收人 ID
fileName	Varchar(50)	空	否		文件名称

表 7.6 JOB 表的结构

字段名称	数据类型	空/非空	是否主键	约束条件	字段描述
jobID	int	非空	是	自增 1	职位 ID
deptID	int	空	否		部门 ID
jobName	varchar(50)	空	否		职位名称
jobText	text	空	否		职位介绍

表 7.7 NOTICE 表的结构

字段名称	数据类型	空/非空	是否主键	约束条件	字段描述
noteID	int	非空	是	自增 1	公告 ID
employeeID	int	空	否		发布人 ID
noteName	varchar(50)	空	否		公告名称
noteInfo	text	空	否		公告介绍
noteTime	varchar(50)	空	否		公告时间

表 7.8 TASK 表的结构

字段名称	数据类型	空/非空	是否主键	约束条件	字段描述
taskID	int	非空	是	自增 1	任务 ID
taskName	Varchar(50)	空	否		任务名称
taskInfo	text	空	否		任务介绍
getTime	datetime	空	否		添加任务时间
taskState	varchar(50)	空	否		任务状态
employeeID	int	空	否		任务添加人 ID

表 7.9 EMPLOYEES 表的结构

字段名称	数据类型	空/非空	是否主键	约束条件	字段描述
employeeID	int	非空	是	自增 1	用户 ID
employeeName	varchar(50)	空	否		用户名字
sex	char(10)	非空	否		用户性别
birthday	DateTime	空	否		用户生日
address	varchar(50)	空	否		地址
NOcode	varchar(50)	空	否		身份证号
learn	varchar(50)	空	否		学历
mobile	varchar(50)	空	否		手机号
email	varchar(50)	空	否		邮箱
jobID	int	空	否		职位 ID
deptID	int	空	否		部门 ID
workState	varchar(50)	空	否		状态
photo	varchar(50)	空	否		照片
agreement	varchar(50)	空	否		就职协议
pwd	char(10)	空	否		密码
userLevel	int	空	否		用户级别
onLine	char(10)	空	否		在线状态

7.8 系统实现与测试

在系统逻辑设计和物理设计完成之后,系统设计即进入系统实现阶段。系统实现是指将系统设计阶段的结果在计算机上实现,将原来图形上的系统方案转换成可执行的应用软件。根据系统设计确立的目标和结构,合理地利用给定的资源,实施可以正常交付运行的实体系统的过程。系统实现是系统开发全过程中费用最大、周期最长、涉及面最广的一个阶段。其中包括开发环境的配置、人员的调配、系统说明书、系统制造基本计划、系统实施计划等,编制制造说明书和制造实施计划,并预测、分析和排除在实施过程中可能出现的随机干扰因素。因此,必须权衡好时间、质量、成本这三者之间的关系,合理安排。

7.8.1 数据库的建立与连接

在 SQL Server 中建立数据库 DatabaseOA,并建立与关系对应的八张表,即 ACTIVE 表、ACTIVEINFO 表、DEPT 表、FILE 表、JOB 表、NOTICE 表、TASK 表、EMPLOYEES 表。

7.8.2 系统实现总框架简介

在数据库和各种表建立之后，在 Microsoft Studio 开发环境中完成控制部分和视图界面部分。

网站实现主要分为 MODEL 层、DAL 层、BLL 层和 Web 层。

7.8.3 系统实现——MODEL 层

Active.cs 中定义了活动类，具体代码如下：

```
using System;
using System.Collections.Generic;
using System.Text;
using System.Data;
using System.Data.OleDb;
namespace MODEL
{
    public class Active
    {
        //构造函数
        public Active(){ }

        ///<summary>
        ///MODEL——活动
        ///</summary>

        #region 字段属性
        private int _activeID;                    //自动编号
        private string _activeName;
        private string _activeInfo;
        private int _activeAgreeNum;

        public int ActiveID
        {
            set { _activeID=value; }
            get { return _activeID; }
        }
        public string ActiveName
        {
            set { _activeName= (value !=null)?value : null; }
            get { return _activeName; }
        }
        public string ActiveInfo
```

```
        {
            set { _activeInfo=(value !=null)?value : null; }
            get { return _activeInfo; }
        }
        public int ActiveAgreeNum
        {
            set { _activeAgreeNum=value; }
            get { return _activeAgreeNum; }
        }
        #endregion
    }
}
```

ActiveInfo.cs 中定义了活动附表类，具体代码如下：

```
using System;
using System.Collections.Generic;
using System.Text;
using System.Data;
using System.Data.OleDb;

namespace MODEL
{
    public class ActiveInfo
    {
        //构造函数
        public ActiveInfo(){ }
        ///<summary>
        ///MODEL——活动附表(存储用户投票信息,防止重复投票)
        ///</summary>
        #region 字段属性
        private int _id;                                    //自动编号
        private int _activeID;
        private int _employeeID;
        public int ID
        {
            set { _id=value; }
            get { return _id; }
        }
        public int ActiveID
        {
            set { _activeID=value; }
            get { return _activeID; }
        }
        public int EmployeeID
```

```
            {
                set { _employeeID=value; }
                get { return _employeeID; }
            }
            #endregion
    }
}
```

Dept.cs 中定义了部门类,具体代码如下:

```
using System;
using System.Collections.Generic;
using System.Text;
using System.Data;
using System.Data.OleDb;

namespace MODEL
{
    public class Dept
    {
        //构造函数
        public Dept(){ }
        ///<summary>
        ///MODEL——部门
        ///</summary>

        #region 字段属性
        private int _deptID;
        private string _deptName;
        private string _deptIntroduce;

        public int DeptID
        {
            set { _deptID=value; }
            get { return _deptID; }
        }
        public string DeptName
        {
            set { _deptName= (value !=null)? value : null; }
            get { return _deptName; }
        }
        public string DeptIntroduce
        {
            set { _deptIntroduce= (value !=null)? value : null; }
            get { return _deptIntroduce; }
```

```
        }
        #endregion
    }
}
```

Emloyee.cs 中定义了员工类,具体代码如下:

```csharp
using System;
using System.Collections.Generic;
using System.Text;
using System.Data;
using System.Data.OleDb;

namespace MODEL
{
    public class Employee
    {
        ///<summary>
        ///MODEL——员工
        ///</summary>

        #region  字段属性
        private int _employeeID;                        //自动编号
        private string _employeeName;
        private string _sex ;
        private DateTime _birthday;
        private string _address;                        //现住址
        private string _no;
        private string _learn;
        private string _mobile;
        private string _email;
        private int _jobID;
        private int _deptID;
        private string _workState;
        private string _photo;
        private string _agreement;
        private string _pwd;
        private string _userLevel;
        private string _onLine;
        public int EmployeeID
        {
            set { _employeeID=value; }
            get { return _employeeID; }
        }
        public string EmployeeName
```

```csharp
{
    set { _employeeName=(value !=null)?value : ""; }
    get { return _employeeName; }
}

public string Sex
{
    set { _sex=(value !=null)?value : null; }
    get { return _sex; }
}
public DateTime Birthday
{
    set { _birthday=value; }
    get { return _birthday; }
}

public string Address
{
    set { _address=(value !=null)?value : null; }
    get { return _address; }
}
public string NO
{
    set { _no=value; }
    get { return _no; }
}
public string Learn
{
    set { _learn=(value !=null)?value : null; }
    get { return _learn; }
}
public string Mobile
{
    set { _mobile=(value !=null)?value : null; }
    get { return _mobile; }
}
public string Email
{
    set { _email=(value !=null)?value : null; }
    get { return _email; }
}
public int JobID
{
    set { _jobID=value; }
```

```
            get { return _jobID; }
        }
        public int DeptID
        {
            set { _deptID=value; }
            get { return _deptID; }
        }
        public string WorkState
        {
            set { _workState=(value !=null)?value : null; }
            get { return _workState; }
        }
        public string Photo
        {
            set { _photo=(value !=null)?value : null; }
            get { return _photo; }
        }
        public string AgreeMent
        {
            set { _agreement=(value !=null)?value : null; }
            get { return _agreement; }
        }
        public string Pwd
        {
            set { _pwd=(value !=null)?value : null; }
            get { return _pwd; }
        }
        public string UserLevel
        {
            set { _userLevel=value; }
            get { return _userLevel; }
        }
        public string OnLine
        {
            set { _onLine=value; }
            get { return _onLine; }
        }
        #endregion
    }
}
```

File.cs 中定义了文件类，具体代码如下：

```
using System;
using System.Collections.Generic;
```

```csharp
using System.Text;
using System.Data;
using System.Data.OleDb;

namespace MODEL
{
    public class File
    {
        //构造函数
        public File(){ }
        ///<summary>
        ///MODEL——文件
        ///</summary>

        #region 字段属性
        private int _fileID;
        private DateTime _fileTime;
        private string _fileText;
        private int _fileFrom;
        private int _fileTo;
        private string _fileName;
        public int FileID
        {
            set { _fileID=value; }
            get { return _fileID; }
        }
        public DateTime FileTime
        {
            set { _fileTime=value; }
            get { return _fileTime; }
        }
        public string FileText
        {
            set { _fileText= (value !=null)?value : null; }
            get { return _fileText; }
        }
        public int FileFrom
        {
            set { _fileFrom=value; }
            get { return _fileFrom; }
        }
        public int FileTo
        {
            set { _fileTo=value; }
```

```
            get { return _fileTo; }
        }
        public string FileName
        {
            set { _fileName= (value !=null)?value : null; }
            get { return _fileName; }
        }
        #endregion
    }
}
```

Job.cs 中定义了工作职位类，具体代码如下：

```
using System;
using System.Collections.Generic;
using System.Text;
using System.Data;
using System.Data.OleDb;

namespace MODEL
{
    public class Job
    {
        ///<summary>
        ///MODEL——工作职位
        ///</summary>
        #region 字段属性
        private int _jobID;
        private int _deptID;
        private string _jobName;
        private string _jobIntroduce;

        public int JobID
        {
            set { _jobID=value; }
            get { return _jobID; }
        }
        public int DeptID
        {
            set { _deptID=value; }
            get { return _deptID; }
        }
        public string JobName
        {
            set { _jobName=value; }
```

```csharp
            get { return _jobName; }
        }
        public string JobIntroduce
        {
            set { _jobIntroduce=value; }
            get { return _jobIntroduce; }
        }
        #endregion
    }
}
```

Notice.cs 中定义了公告类，具体代码如下：

```csharp
using System;
using System.Collections.Generic;
using System.Text;
using System.Data;
using System.Data.OleDb;

namespace MODEL
{
    public class Notice
    {
        //构造函数
        public Notice(){ }

        ///<summary>
        ///MODEL—公告
        ///</summary>

        #region 字段属性
        private int _noteID;                                //自动编号
        private int _employeeID;
        private string _noteName;
        private string _noteInfo;
        private string  _noteTime;

        public int NoteID
        {
            set { _noteID=value; }
            get { return _noteID; }
        }
        public int EmployeeID
        {
            set { _employeeID=value; }
```

```csharp
            get { return _employeeID; }
        }
        public string NoteName
        {
            set { _noteName=(value !=null)?value : null; }
            get { return _noteName; }
        }
        public string NoteInfo
        {
            set { _noteInfo=(value !=null)?value : null; }
            get { return _noteInfo; }
        }
        public string NoteTime
        {
            set { _noteTime=value; }
            get { return _noteTime; }
        }
        #endregion
    }
}
```

Task.cs 中定义了工作任务类,具体代码如下:

```csharp
using System;
using System.Collections.Generic;
using System.Text;
using System.Data;
using System.Data.OleDb;

namespace MODEL
{
    public class Task
    {
        //构造函数
        public Task(){ }
        ///<summary>
        ///MODEL——工作任务
        ///</summary>
        #region 字段属性
        private int _taskID;
        private string _taskName;
        private string _taskInfo;
        private DateTime _getTime;
        private string _tastState;
        private int employeeID;
```

```csharp
        public int EmployeeID
        {
            get { return employeeID; }
            set { employeeID=value; }
        }
        public int TaskID
        {
            set { _taskID=value; }
            get { return _taskID; }
        }
        public string TaskName
        {
            set { _taskName=(value !=null)?value : null; }
            get { return _taskName; }
        }
        public string TaskInfo
        {
            set { _taskInfo=(value !=null)?value : null; }
            get { return _taskInfo; }
        }
        public DateTime GetTime
        {
            set { _getTime=value; }
            get { return _getTime; }
        }
        public string TaskState
        {
            set { _tastState=(value !=null)?value : null; }
            get { return _tastState; }
        }
        #endregion
    }
}
```

7.8.4 系统实现——DAL 层

DbConn.cs 数据库连接类,具体代码如下:

```csharp
using System;
using System.Collections.Generic;
using System.Text;
using System.Configuration;
using System.Data;
using System.Data.SqlClient;
```

```csharp
namespace DAL
{
    public class DbConn
    {
        ///<summary>
        ///数据库连接类
        ///</summary>
        public static SqlConnection conn=null;
        public static string connectString = ConfigurationManager.ConnectionStrings["conn"].ToString();
        //构造函数
        public static SqlConnection getConn()
        {
            return conn=new SqlConnection(connectString);
        }
        //静态返回连接字符串
        public static string GetConnectionString()
        {
            return connectString;
        }
        //获取数据集或数据
        public static DataSet getData(string sqlString)
        {
            SqlConnection conn=getConn();
            conn.Open();
            SqlDataAdapter oda=new SqlDataAdapter(sqlString, conn);
            DataSet ds=new DataSet();
            oda.Fill(ds);

            conn.Close();
            return ds;
        }
        public static DataSet getData(string sqlString, params SqlParameter[] SqlParams)
        {
            SqlConnection conn=getConn();
            conn.Open();
            SqlDataAdapter oda=new SqlDataAdapter(sqlString, conn);
            DataSet ds=new DataSet();
            if(SqlParams !=null)
            {
                foreach(SqlParameter p in SqlParams)
                {
                    oda.SelectCommand.Parameters.Add(p);
```

```csharp
            }
            oda.Fill(ds);
        }
        conn.Close();
        return ds;
    }
    public static SqlDataReader getReader(string sqlString)
    {
        SqlConnection conn=getConn();
        conn.Open();
        SqlCommand cmd=new SqlCommand(sqlString, conn);
        SqlDataReader reader=cmd.ExecuteReader(CommandBehavior.CloseConnection);
        return reader;
    }
    public static SqlDataReader getReader ( string sqlString, params SqlParameter[] SqlParams)
    {
        SqlConnection conn=getConn();
        conn.Open();
        SqlCommand cmd=new SqlCommand(sqlString, conn);
        SqlDataReader reader=null;
        if(SqlParams !=null)
        {
            foreach(SqlParameter p in SqlParams)
            {
                cmd.Parameters.Add(p);
            }
            reader=cmd.ExecuteReader(CommandBehavior.CloseConnection);
        }
        return reader;
    }
  }
}
//////返回 DataReader 对象时要注意关闭数据库
```

DALactive.cs 是活动类的方法实现,具体代码如下：

```csharp
using System;
using System.Collections.Generic;
using System.Text;
using System.Data;
using System.Data.SqlClient;
namespace DAL
{
    public class DALactive
```

```csharp
{
    ///<summary>
    ///活动
    ///</summary>
    ///<param name="active"></param>

    #region 插入
    public void Insert(MODEL.Active active)
    {
        string sql = @" insert into ACTIVE ( activeName, activeInfo,
        activeAgreeNum)values(@activeName,@activeInfo,@activeAgreeNum)";
        SqlParameter[] SqlParams={
            new SqlParameter("@activeName",active.ActiveName),
            new SqlParameter("@activeInfo",active.ActiveInfo),
            new SqlParameter("@activeAgreeNum",active.ActiveAgreeNum)
        };
        SqlHelper. ExecuteNonQuery ( DbConn. GetConnectionString ( ), sql,
        SqlParams);
    }
    #endregion

    #region 更新
    public void Update(MODEL.Active active)
    {
        string sql="update ACTIVE set activeName=@activeName,activeInfo=
        @activeInfo, activeAgreeNum = @ activeAgreeNum where activeID =
        @activeID";
        SqlParameter[] SqlParams={
            new SqlParameter("@activeName",active.ActiveName),
            new SqlParameter("@activeInfo",active.ActiveInfo),
            new SqlParameter("@activeAgreeNum",active.ActiveAgreeNum),
            new SqlParameter("@activeID",active.ActiveID)
        };
        SqlHelper. ExecuteNonQuery ( DbConn. GetConnectionString ( ), sql,
        SqlParams);
    }
    #endregion

    #region 删除
    public void Delete(MODEL.Active active)
    {
        string sql="delete from ACTIVE where activeID=@activeID";
        SqlParameter[] SqlParams= { new SqlParameter ("@activeID", active.
```

```csharp
            ActiveID)};
            SqlHelper.ExecuteNonQuery(DbConn.GetConnectionString(), sql,
            SqlParams);
        }
        #endregion

        public MODEL.Active getActiveModel(int id)
        {
            string sql="select * from ACTIVE where activeID=@id";
            SqlParameter[] sp={
            new SqlParameter("@id",id)
            };
            MODEL.Active active=new MODEL.Active();
            SqlDataReader read=DbConn.getReader(sql, sp);
            if(read.Read())
            {
                active.ActiveAgreeNum=Convert.ToInt32(read["activeAgreeNum"]);
                active.ActiveID=id;
                active.ActiveInfo=read["activeInfo"].ToString();
                active.ActiveName=read["activeName"].ToString();
            }
            read.Close();
            return active;
        }
    }
}
```

DALactiveInfo.cs 中的具体代码如下：

```csharp
using System;
using System.Collections.Generic;
using System.Text;
using System.Data;
using System.Data.SqlClient;
namespace DAL
{
    public class DALactiveInfo
    {
        ///<summary>
        ///MODEL——操作活动附表(存储用户投票信息,防止重复投票)
        ///</summary>

        #region 添加
        public void Insert(int actID, int empID)
        {
```

```csharp
    string sql =" insert into ACTIVEINFO (activeID, employeeID) values
    (@activeID,@employeeID)";
    SqlParameter[] SqlParams={
        new SqlParameter("@activeID",actID),
        new SqlParameter("@employeeID",empID)
    };
    SqlHelper.ExecuteNonQuery(DbConn.GetConnectionString(), sql, SqlParams);
}

#endregion

#region 更新
public void Update(MODEL.ActiveInfo activeInfo)
{
    string sql="update ACTIVEINFO set activeID=@activeID,employeeID=
    @employeeID where ID=@ID";
    SqlParameter[] SqlParams={
        new SqlParameter("@activeID",activeInfo.ActiveID),
        new SqlParameter("@employeeID",activeInfo.EmployeeID),
        new SqlParameter("@ID",activeInfo.ID)
    };
    SqlHelper.ExecuteNonQuery(DbConn.GetConnectionString(), sql, SqlParams);
}
#endregion

#region 删除
public void Delete(MODEL.ActiveInfo activeInfo)
{
    string sql="delete from ACTIVEINFO where ID=@ID";
    SqlParameter[] SqlParams={
    new SqlParameter("@ID",activeInfo.ID)
    };
    SqlHelper.ExecuteNonQuery(DbConn.GetConnectionString(), sql, SqlParams);
}
#endregion

#region 验证
public bool checkActive(int actID, int empID)
{
    string sql="select count(*) from ACTIVEINFO where activeID=@actID
    and employeeID=@id";
    SqlParameter[] sp={
        new SqlParameter("@actID",actID),
        new SqlParameter("@id",empID)
```

```
            };
            int i = SqlHelper.ExecuteQuery(DbConn.GetConnectionString(), sql,
            sp);
            if(i>0)
            {
                return false;
            }
            else
            {
                return true;
            }
        }
        #endregion
    }
}
```

DALdept.cs 中的具体代码如下:

```
using System;
using System.Collections.Generic;
using System.Text;
using System.Data;
using System.Data.SqlClient;
namespace DAL
{
    public class DALdept
    {
        ///<summary>
        ///部门
        ///</summary>
        #region 添加
        public void Insert(MODEL.Dept dept)
        {
            string sql="insert into DEPT(deptName,deptText)values(@deptName,@deptText)";
            SqlParameter[] SqlParams={
                new SqlParameter("@deptName",dept.DeptName),
                new SqlParameter("@deptText",dept.DeptIntroduce)
            };
            SqlHelper.ExecuteNonQuery(DbConn.GetConnectionString(), sql, SqlParams);
        }
        #endregion

        #region 更新
```

```csharp
public void Update(MODEL.Dept dept)
{
    string sql="update DEPT set deptName=@deptName,deptText=@deptText
    where deptID=@deptID";
    SqlParameter[] SqlParams={
        new SqlParameter("@deptName",dept.DeptName),
        new SqlParameter("@deptText",dept.DeptIntroduce),
            new SqlParameter("@deptID",dept.DeptID)
    };
    SqlHelper.ExecuteNonQuery(DbConn.GetConnectionString(), sql, SqlParams);
}
#endregion

#region 删除
public void Delete(MODEL.Dept dept)
{
    string sql="delete from DEPT where deptID=@deptID";
    SqlParameter[] SqlParams={
        new SqlParameter("@deptID",dept.DeptID)
    };
    SqlHelper.ExecuteNonQuery(DbConn.GetConnectionString(), sql, SqlParams);
}
#endregion
#region 查询
public DataSet getDept()
{
    string sql="select * from DEPT";
    return DbConn.getData(sql);
}
public MODEL.Dept getDeptModel(int id)
{
    string sql="select deptName,deptText from DEPT where deptID=@id";
    SqlParameter[] sp={
    new SqlParameter("@id",id)
    };
    SqlDataReader read=DbConn.getReader(sql, sp);
    MODEL.Dept dept=new MODEL.Dept();
    if(read.Read())
    {
        dept.DeptID=id;
        dept.DeptName=read["deptName"].ToString();
        dept.DeptIntroduce=read["deptText"].ToString();
    }
    read.Close();
```

```csharp
            return dept;
        }
        #endregion

        #region 验证
        public bool checkDept(string name)
        {
            string sql="select count(*) from DEPT where deptName=@name";
            SqlParameter[] sqlParaeter={ new SqlParameter("@name", name)};
            int i=   SqlHelper.ExecuteQuery(DbConn.GetConnectionString(), sql, sqlParaeter);
            if(i>0)
            {
                return false;                      //验证不通过
            }
            else
            {
                return true;                       //通过验证
            }
        }
        #endregion
    }
}
```

DALemployee.cs 中的具体代码如下：

```csharp
using System;
using System.Collections.Generic;
using System.Text;
using System.Data;
using System.Data.SqlClient;
using MODEL;
namespace DAL
{
    public class DALemployee
    {
        #region 验证
        ///<summary>
        ///验证用户密码
        ///</summary>
        ///<param name="user"></param>
        ///<param name="password"></param>
        ///<returns></returns>

        public string loginUserLevel(string user,string password)
```

```csharp
        {
            string sql="select userLevel from EMPLOYEES where employeeName=@
            user and pwd=@pwd";
            SqlParameter[] para={
                        new SqlParameter("@user",user),
                        new SqlParameter("@pwd",password)
            };

            SqlDataReader reader=DbConn.getReader(sql, para);
            if(reader.Read())
            {
                return reader["userLevel"].ToString();
            }
            else
            {
                return "0";
            }
        }
        ///<summary>
        ///用户登录权限
        ///</summary>
        ///<param name="userName"></param>
        ///<returns></returns>
        public string loginUserLevel(string userName)
        {
            string sql="select userLevel from EMPLOYEES where employeeName=@
            user";
            SqlParameter[] para={
                        new SqlParameter("@user",userName),
            };
            SqlDataReader reader=DbConn.getReader(sql, para);
            if(reader.Read())
            {
                return reader["userLevel"].ToString();
            }
            else
            {
                return "0";
            }
        }
        #endregion
        //添加,删除,修改
        #region 插入
        public void Insert(MODEL.Employee emp)
```

```csharp
{
    string sql = @" insert into EMPLOYEES (employeeName, sex, birthday,
    address, NOcode, learn, mobile, email, jobID, deptID, workState, photo,
    agreement, userLevel)
        values(@employeeName,@sex,@birthday,@address,@NO,@learn,
        @mobile,@email,@jobID,@deptID,@workState,@photo,@agreeMent,
        @userLevel)";
    SqlParameter[] SqlParams={
        new SqlParameter("@employeeName",emp.EmployeeName),
        new SqlParameter("@sex",emp.Sex),
        new SqlParameter("@birthday",emp.Birthday),
        new SqlParameter("@address",emp.Address),
        new SqlParameter("@NO",emp.NO),
        new SqlParameter("@learn",emp.Learn),
        new SqlParameter("@mobile",emp.Mobile),
        new SqlParameter("@email",emp.Email),
        new SqlParameter("@jobID",emp.JobID),
        new SqlParameter("@deptID",emp.DeptID),
        new SqlParameter("@workState",emp.WorkState),
        new SqlParameter("@photo",emp.Photo),
        new SqlParameter("@agreeMent",emp.AgreeMent),
        new SqlParameter("@userLevel",emp.UserLevel)
    };
    SqlHelper.ExecuteNonQuery(DbConn.GetConnectionString(), sql, SqlParams);
}
#endregion
#region 更新
public void Update(MODEL.Employee emp)
{
    string sql=@"update EMPLOYEES set employeeName=@employeeName,sex=
    @sex,birthday=@birthday,address=@address,
        NOcode=@NO,learn=@learn,mobile=@mobile,email=@email,
        jobID=@jobID,deptID=@deptID,workState=@workState,
        photo=@photo,agreement=@agreeMent,pwd=@pwd,userLevel=
        @userLevel where employeeID=@employeeID";
    SqlParameter[] SqlParams={
        new SqlParameter("@employeeName",emp.EmployeeName),
        new SqlParameter("@sex",emp.Sex),
        new SqlParameter("@birthday",emp.Birthday),
        new SqlParameter("@address",emp.Address),
        new SqlParameter("@NO",emp.NO),
        new SqlParameter("@learn",emp.Learn),
        new SqlParameter("@mobile",emp.Mobile),
        new SqlParameter("@email",emp.Email),
```

```csharp
                new SqlParameter("@jobID",emp.JobID),
                new SqlParameter("@deptID",emp.DeptID),
                new SqlParameter("@workState",emp.WorkState),
                new SqlParameter("@photo",emp.Photo),
                new SqlParameter("@agreeMent",emp.AgreeMent),
                new SqlParameter("@pwd",emp.Pwd),
                new SqlParameter("@userLevel",emp.UserLevel),
                new SqlParameter("@employeeID",emp.EmployeeID)
            };
            SqlHelper.ExecuteNonQuery(DbConn.GetConnectionString(), sql, SqlParams);
        }
        #endregion

        #region 删除
        public void Delete(MODEL.Employee emp)
        {
            string sql="delete from EMPLOYEES where employeeID=@employeeID";
            SqlParameter[] SqlParams = { new SqlParameter ("@employeeID", emp.EmployeeID)};
            SqlHelper.ExecuteNonQuery(DbConn.GetConnectionString(), sql, SqlParams);
        }
        #endregion
        ///<summary>
        ///这个是为了方便绑定数据
        ///</summary>
        ///<param name="empName"></param>
        ///<returns></returns>
        #region select
        public DataSet getEmp(string empName)
        {
            string sql = @"select * from EMPLOYEES, DEPT, JOB WHERE EMPLOYEES.deptID=DEPT.deptID and EMPLOYEES.jobID=JOB.jobID and employeeName=@name";
            SqlParameter[] Sqlparams={
                new SqlParameter("@name",empName)
            };
            return DbConn.getData(sql, Sqlparams);
        }
        public DataSet getEmp(int id)
        {
            string sql = @"select * from EMPLOYEES, DEPT, JOB WHERE EMPLOYEES.deptID=DEPT.deptID and EMPLOYEES.jobID=JOB.jobID and employeeID=
```

```csharp
        @employeeID";
        SqlParameter[] Sqlparams={
            new SqlParameter("@employeeID",id)
        };
        return DbConn.getData(sql,Sqlparams);
}
///<summary>
///获取在线用户信息
///</summary>
///<param name="online"></param>
///<returns></returns>
public DataSet OnLine()
{
        string sql=@"select employeeID, employeeName, deptName, jobName,
        photo from EMPLOYEES ,DEPT,JOB WHERE EMPLOYEES.deptID=DEPT.deptID
        and EMPLOYEES.jobID=JOB.jobID and onLine='1'";
        return DbConn.getData(sql);
}
///<summary>
///获取员工实体 重载两个方法
///</summary>
///<param name="id"></param>
///<returns></returns>
public MODEL.Employee getModel(int id)
{
        string sql=@"select [employeeName],[sex],[birthday],[address],
        [NOcode],[learn],[mobile],[email],[jobID],[deptID],[workState],
        [photo],[agreement],[pwd],[userLevel],[onLine]  from EMPLOYEES
        where employeeID=@id";
        SqlParameter[] Sqlparams={
            new SqlParameter("@id",id)
        };
        MODEL.Employee empModel=new Employee();
        SqlDataReader read=DbConn.getReader(sql,Sqlparams);
        if(read.Read())
        {
            empModel.Address=read["address"].ToString();
            empModel.AgreeMent=read["agreement"].ToString();
            empModel.Birthday = Convert.ToDateTime(read["birthday"]==
            DBNull.Value ?null : read["birthday"]);
            empModel.DeptID=Convert.ToInt32(read["deptID"]);
            empModel.JobID=Convert.ToInt32(read["jobID"]);
            empModel.Mobile=read["mobile"].ToString();
            empModel.NO=read["NOcode"].ToString();
```

```csharp
            empModel.OnLine=read["onLine"].ToString();
            empModel.Photo=read["photo"].ToString();
            empModel.Pwd=read["pwd"].ToString();
            empModel.Sex=read["sex"].ToString();
            empModel.UserLevel=read["userLevel"].ToString();
            empModel.WorkState=read["workState"].ToString();
            empModel.EmployeeName=read["employeeName"].ToString();
            empModel.Learn=read["learn"].ToString();
            empModel.Email=read["email"].ToString();
            empModel.EmployeeID=id;
        }
        read.Close();
        return empModel;
    }
    public MODEL.Employee getModel(string name)
    {
        string sql = @" select [employeeID],[sex],[birthday],[address],
        [NOcode],[learn],[mobile],[email],[jobID],[deptID],[workState],
        [photo],[agreement],[pwd],[userLevel],[onLine]
            from EMPLOYEES where employeeName=@name";
        SqlParameter[] Sqlparams={
          new SqlParameter("@name",name)
        };
        MODEL.Employee empModel=new Employee();
        SqlDataReader read=DbConn.getReader(sql, Sqlparams);
        if(read.Read())
        {
            empModel.Address=read["address"].ToString();
            empModel.AgreeMent=read["agreement"].ToString();
            empModel.Birthday = Convert.ToDateTime(read["birthday"]==
            DBNull.Value?null:read["birthday"]);
            empModel.DeptID=Convert.ToInt32(read["deptID"]);
            empModel.JobID=Convert.ToInt32(read["jobID"]);
            empModel.Mobile=read["mobile"].ToString();
            empModel.NO=read["NOcode"].ToString();
            empModel.OnLine=read["onLine"].ToString();
            empModel.Photo=read["photo"].ToString();
            empModel.Pwd=read["pwd"].ToString();
            empModel.Sex=read["sex"].ToString();
            empModel.UserLevel=read["userLevel"].ToString();
            empModel.WorkState=read["workState"].ToString();
            empModel.EmployeeName=name;
            empModel.Learn=read["learn"].ToString();
            empModel.Email=read["email"].ToString();
```

```csharp
            empModel.EmployeeID=Convert.ToInt32(read["employeeID"]);
        }
        read.Close();
        return empModel;
    }
    #endregion

    ///<summary>
    ///更新在线状态
    ///</summary>
    ///<param name="Online"></param>
    ///<param name="userName"></param>
    public void getOnline(string Online, string userName)
    {
        string sql =" update EMPLOYEES set onLine = '" + Online +" ' where employeeName='"+userName+"'";
        SqlHelper.ExecuteNonQuery(DbConn.GetConnectionString(), sql);
    }
 }
}
```

DALfile.cs 里的具体代码如下：

```csharp
using System;
using System.Collections.Generic;
using System.Text;
using System.Data;
using System.Data.SqlClient;
namespace DAL
{
    public class DALfile
    {
        ///<summary>
        ///文件
        ///</summary>
        #region 添加
        public void Insert(MODEL.File getfile)
        {
            string sql=@"insert into [FILE](fileTime,fileText,fileFrom,fileTo,fileName) values (@fileTime, @fileText, @fileFrom, @fileTo, @fileName)";
            SqlParameter[] SqlParams={
                new SqlParameter("@fileTime",getfile.FileTime),
                new SqlParameter("@fileText",getfile.FileText),
```

```csharp
            new SqlParameter("@fileFrom",getfile.FileFrom),
            new SqlParameter("@fileTo",getfile.FileTo),
            new SqlParameter("@fileName",getfile.FileName)
        };
        SqlHelper.ExecuteNonQuery(DbConn.GetConnectionString(), sql, SqlParams);
    }
    #endregion

    #region 更新
    public void Update(MODEL.File getfile)
    {
        string sql = @"update [FILE] set fileTime=@fileTime, fileText=
        @fileText ,fileFrom=@fileFrom,fileTo=@fileTo,fileName=@fileName
        where fileID=@fileID";
        SqlParameter[] SqlParams={
            new SqlParameter("@fileTime",getfile.FileTime),
            new SqlParameter("@fileText",getfile.FileText),
            new SqlParameter("@fileFrom",getfile.FileFrom),
            new SqlParameter("@fileTo",getfile.FileTo),
            new SqlParameter("@fileName",getfile.FileName),
            new SqlParameter("@fileID",getfile.FileID)
        };
        SqlHelper.ExecuteNonQuery(DbConn.GetConnectionString(), sql, SqlParams);
    }
    #endregion

    #region 删除
    public void Delete(MODEL.File getfile)
    {
        string sql="delete from [FILE] where fileID=@fileID";
        SqlParameter[] SqlParams={
            new SqlParameter("@fileID",getfile.FileID)
        };
        SqlHelper. ExecuteNonQuery (DbConn. GetConnectionString ( ), sql,
        SqlParams);
    }
    #endregion
    ///<summary>
    ///文件实体
    ///</summary>
    ///<param name="id"></param>
    ///<returns></returns>
    public MODEL.File getFileModel(int id)
    {
```

```csharp
string sql="select * from [FILE] where fileID=@id";
SqlParameter[] sp={
    new SqlParameter("@id",id)
};
MODEL.File files=new MODEL.File();
SqlDataReader read=DbConn.getReader(sql, sp);
if(read.Read())
{
    files.FileID=id;
    files.FileName=read["fileName"].ToString();
    files.FileText=read["fileText"].ToString();
    files.FileTime=Convert.ToDateTime(read["fileTime"]);
    files.FileFrom=Convert.ToInt32(read["fileFrom"]);
    files.FileTo=Convert.ToInt32(read["fileTo"]);
}
read.Close();
return files;
        }
    }
}
```

DALjob.cs 里具体代码如下：

```csharp
using System;
using System.Collections.Generic;
using System.Text;
using System.Data;
using System.Data.SqlClient;
namespace DAL
{
    public class DALjob
    {
        ///<summary>
        ///职位
        ///</summary>
        #region 添加
        public void Insert(MODEL.Job job)
        {
            string sql="insert into JOB(deptID,jobName,jobText)values(@deptID,
            @jobName,@jobText)";
            SqlParameter[] SqlParams={
                new SqlParameter("@deptID",job.DeptID),
                new SqlParameter("@jobName",job.JobName),
                new SqlParameter("@jobText",job.JobIntroduce)
            };
```

```csharp
            SqlHelper.ExecuteNonQuery(DbConn.GetConnectionString(), sql, SqlParams);
}
#endregion

#region 更新
public void Update(MODEL.Job job)
{
    string sql =" update JOB set deptID=@deptID,jobName=@jobName,
    jobText=@jobText where jobID=@jobID";
    SqlParameter[] SqlParams={
        new SqlParameter("@deptID",job.DeptID),
        new SqlParameter("@jobName",job.JobName),
        new SqlParameter("@jobText",job.JobIntroduce),
        new SqlParameter("@jobID",job.JobID)
    };
    SqlHelper.ExecuteNonQuery(DbConn.GetConnectionString(), sql, SqlParams);
}
#endregion

#region 删除
public void Delete(MODEL.Job job)
{
    string sql="delete from JOB where jobID=@jobID";
    SqlParameter[] SqlParams={
        new SqlParameter("@jobID",job.JobID)
    };
    SqlHelper.ExecuteNonQuery(DbConn.GetConnectionString(), sql, SqlParams);
}
#endregion

#region 查询
public DataSet getJobName(int id)
{
    string sql="select jobID,jobName from JOB where deptID="+id;
    return DbConn.getData(sql);
}

public MODEL.Job getDeptModel(int id)
{
    string sql="select * from JOB where jobID=@id";
    SqlParameter[] sp={
    new SqlParameter("@id",id)
```

```csharp
            };
            MODEL.Job job=new MODEL.Job();
            SqlDataReader read=DbConn.getReader(sql, sp);
            if(read.Read())
            {
                job.JobID=id;
                job.DeptID=Convert.ToInt32(read["deptID"]);
                job.JobName=read["jobName"].ToString();
                job.JobIntroduce=read["jobText"].ToString();

            }
            read.Close();
            return job;

        }
#endregion

        #region  验证
        public bool checkJob(string name)
        {
            string sql="select count(*) from JOB where jobName=@name";
            SqlParameter[] sqlParaeter={ new SqlParameter("@name", name)};
            int i = SqlHelper.ExecuteQuery(DbConn.GetConnectionString(), sql,
            sqlParaeter);
            if(i>0)
            {
                return false;
            }
            else
            {
                return true;
            }
        }
        #endregion
    }
}
```

DALnotice.cs 中具体代码如下：

```csharp
using System;
using System.Collections.Generic;
using System.Text;
using System.Data;
using System.Data.SqlClient;
namespace DAL
```

```csharp
{
    public class DALnotice
    {
        ///<summary>
        ///公告
        ///</summary>
        #region 添加
        public void Insert(MODEL.Notice notice)
        {
            string sql = @" insert into NOTICE (employeeID, noteName, noteInfo, noteTime)
                            values(@employeeID,@noteName,@noteInfo,@noteTime)";
            SqlParameter[] SqlParams={
                new SqlParameter("@employeeID",notice.EmployeeID),
                new SqlParameter("@noteName",notice.NoteName),
                new SqlParameter("@noteInfo",notice.NoteInfo),
                new SqlParameter("@noteTime",notice.NoteTime)
            };
            SqlHelper.ExecuteNonQuery(DbConn.GetConnectionString(), sql, SqlParams);
        }
        #endregion

        #region 更新
        public void Update(MODEL.Notice notice)
        {
            string sql=@"update NOTICE set employeeID=@employeeID, noteName=
                @noteName,noteInfo=@noteInfo  where noteID=@noteID";
            SqlParameter[] SqlParams={
                new SqlParameter("@employeeID",notice.EmployeeID),
                new SqlParameter("@noteName",notice.NoteName),
                new SqlParameter("@noteInfo",notice.NoteInfo),
                new SqlParameter("@noteID",notice.NoteID)
            };
            SqlHelper.ExecuteNonQuery(DbConn.GetConnectionString(), sql, SqlParams);
        }
        #endregion

        #region 删除
        public void Delete(MODEL.Notice notice)
        {
            string sql="delete from NOTICE where noteID=@noteID";
            SqlParameter[] SqlParams={
                new SqlParameter("@noteID",notice.NoteID)
            };
```

```csharp
            SqlHelper.ExecuteNonQuery(DbConn.GetConnectionString(), sql, SqlParams);
        }
        #endregion
        public MODEL.Notice getNoteModel(int id)
        {
            string sql="select * from NOTICE where noteID=@noteID";
            SqlParameter[] SqlParams={ new SqlParameter("@noteID", id)};
            MODEL.Notice notice=new MODEL.Notice();
            SqlDataReader read=DbConn.getReader(sql, SqlParams);
            if(read.Read())
            {
                notice.EmployeeID=Convert.ToInt32(read["employeeID"]);
                notice.NoteID=id;
                notice.NoteName=read["noteName"].ToString();
                notice.NoteTime=read["noteTime"].ToString();
                notice.NoteInfo=read["noteInfo"].ToString();
            }
            read.Close();
            return notice;
        }
    }
}
```

DALtask.cs 中的具体代码如下：

```csharp
using System;
using System.Collections.Generic;
using System.Text;
using System.Data;
using System.Data.SqlClient;
namespace DAL
{
    public class DALtask
    {
        ///<summary>
        ///工作任务
        ///</summary>
        #region 添加
        public void Insert(MODEL.Task task)
        {
            string sql=@"insert into TASK(taskName,taskInfo,getTime, taskState,
                employeeID)
values(@taskName,@taskInfo,@getTime,@taskState,@employeeID)";
            SqlParameter[] SqlParams={
                new SqlParameter("@taskName",task.TaskName),
```

```csharp
        new SqlParameter("@taskInfo",task.TaskInfo),
        new SqlParameter("@getTime",task.GetTime),
        new SqlParameter("@taskState",task.TaskState),
        new SqlParameter("@employeeID",task.EmployeeID)
    };
    SqlHelper.ExecuteNonQuery(DbConn.GetConnectionString(), sql, SqlParams);
}
#endregion

#region 更新
public void Update(MODEL.Task task)
{
    string sql=@"update TASK set taskName=@taskName, taskInfo=
    @taskInfo,getTime=@getTime, taskState=@taskState where taskID=
    @taskID";
    SqlParameter[] SqlParams={
        new SqlParameter("@taskName",task.TaskName),
        new SqlParameter("@taskInfo",task.TaskInfo),
        new SqlParameter("@getTime",task.GetTime),
        new SqlParameter("@taskState",task.TaskState),
        new SqlParameter("@taskID",task.TaskID)
    };
    SqlHelper.ExecuteNonQuery(DbConn.GetConnectionString(), sql, SqlParams);
}
#endregion

#region 删除
public void Delete(MODEL.Task task)
{
    string sql="delete from TASK where taskID=@taskID";
    SqlParameter[] SqlParams={
        new SqlParameter("@taskID",task.TaskID)
    };
    SqlHelper.ExecuteNonQuery(DbConn.GetConnectionString(), sql, SqlParams);
}
#endregion
#region 验证
public bool checkTask(string name)
{
    string sql="select count(*) from TASK where taskName=@name";
    SqlParameter[] sqlParaeter={ new SqlParameter("@name", name) };
    int i = SqlHelper.ExecuteQuery(DbConn.GetConnectionString(), sql, sqlParaeter);
    if(i>0)
```

```
            {
                return false;
            }
            else
            {
                return true;
            }
        }
        #endregion
    }
}
```

7.8.5 系统实现——BLL 层

BLLactive.cs 中的具体代码如下:

```
using System;
using System.Collections.Generic;
using System.Text;

namespace BLL
{
    public class BLLactive
    {
        private static readonly  DAL.DALactive getActive=new DAL.DALactive();

        #region 插入
        public static void Insert(MODEL.Active active)
        {
            getActive.Insert(active);
        }
        #endregion

        #region 更新
        public static void Update(MODEL.Active active)
        {
            getActive.Update(active);
        }
        #endregion

        #region 删除
        public static void Delete(MODEL.Active active)
        {
```

```
            getActive.Delete(active);
        }
        #endregion
    }
}
```

BLLactiveInfo.cs 中的具体代码如下：

```
using System;
using System.Collections.Generic;
using System.Text;

namespace BLL
{
    public class BLLactiveInfo
    {
        private static readonly DAL.DALactiveInfo getActiveInfo = new DAL.DALactiveInfo();
        ///<summary>
        ///MODEL——操作活动附表（存储用户投票信息，防止重复投票）
        ///</summary>

        #region 添加
        public static void Insert(int actID, int empID)
        {
            getActiveInfo.Insert(actID, empID);
        }
        #endregion

        #region 更新
        public static void Update(MODEL.ActiveInfo activeInfo)
        {
            getActiveInfo.Update(activeInfo);
        }
        #endregion

        #region 删除
        public static void Delete(MODEL.ActiveInfo activeInfo)
        {
            getActiveInfo.Delete(activeInfo);
        }
        #endregion
```

```csharp
        public static bool checkActive(int actID,int empID)
        {
            return getActiveInfo.checkActive(actID,empID);
        }
    }
}
```

BLLdept.cs 里的具体代码如下：

```csharp
using System;
using System.Collections.Generic;
using System.Text;
using System.Data;
using System.Data.OleDb;
namespace BLL
{
    public class BLLdept
    {
        private static readonly DAL.DALdept getDept=new DAL.DALdept();

        ///<summary>
        ///部门
        ///</summary>
        #region 添加
        public static   void Insert(MODEL.Dept dept)
        {
            getDept.Insert(dept);
        }
        #endregion

        #region 更新
        public static   void Update(MODEL.Dept dept)
        {
            getDept.Update(dept);
        }
        #endregion

        #region 删除
        public static   void Delete(MODEL.Dept dept)
        {
            getDept.Delete(dept);

        }
        #endregion
```

```csharp
            #region 查询
            public static DataSet Dept()
            {
                return getDept.getDept();
            }

            public static MODEL.Dept getDeptModel(int id)
            {
                return getDept.getDeptModel(id);
            }
            #endregion

            #region 验证
            public static bool checkDept(string name)
            {
                return getDept.checkDept(name);
            }
            #endregion
        }
    }
```

BLLemployee.cs 中的具体代码如下：

```csharp
using System;
using System.Collections.Generic;
using System.Text;
using System.Data;
using System.Data.SqlClient;
namespace BLL
{
    public class BLLemployee
    {
        private static readonly DAL.DALemployee getEmployee=new DAL.DALemployee();

        #region 插入
        public static void Insert(MODEL.Employee emp)
        {
            getEmployee.Insert(emp);
        }
        #endregion


        #region 更新
```

```
public static void Update(MODEL.Employee emp)
{

    getEmployee.Update(emp);
}
#endregion

#region 删除
public void Delete(MODEL.Employee emp)
{
    getEmployee.Delete(emp);
}
#endregion
#region select
public static DataSet getEmp(string empName)
{
    return getEmployee.getEmp(empName);
}
public static DataSet getEmp(int id)
{
    return getEmployee.getEmp(id);
}
public static DataSet OnLine()
{
    return getEmployee.OnLine();
}
///<summary>
///重载两个方法到员工 MODEL
///</summary>
///<param name="id"></param>
///<returns></returns>
public static MODEL.Employee getMobile(int id)
{
    return getEmployee.getModel(id);
}
public static MODEL.Employee getMobile(string name)
{
    return getEmployee.getModel(name);
}
#endregion
    }
}
```

BLLfile.cs 中的具体代码如下：

```csharp
using System;
using System.Collections.Generic;
using System.Text;

namespace BLL
{
    public class BLLfile
    {
        private static readonly DAL.DALfile getFile=new DAL.DALfile();
        ///<summary>
        ///文件
        ///</summary>
        #region 添加
        public static  void Insert(MODEL.File getfile)
        {
            getFile.Insert(getfile);
        }
        #endregion
        #region 更新
        public static  void Update(MODEL.File getfile)
        {
            getFile.Update(getfile);
        }
        #endregion
        #region 删除
        public static  void Delete(MODEL.File getfile)
        {
            getFile.Delete(getfile);
        }
        #endregion
        public static MODEL.File getFileModel(int id)
        {
            return getFile.getFileModel(id);
        }
    }
}
```

BLLjob.cs 中的具体代码如下：

```csharp
using System;
using System.Collections.Generic;
using System.Text;
using System.Data;
using System.Data.OleDb;
namespace BLL
```

```csharp
{
    public class BLLjob
    {
        public static readonly DAL.DALjob getJob=new DAL.DALjob();
        ///<summary>
        ///职位
        ///</summary>
        #region 添加
        public static  void Insert(MODEL.Job job)
        {
            getJob.Insert(job);
        }
        #endregion
        #region 更新
        public static void Update(MODEL.Job job)
        {
            getJob.Update(job);
        }
        #endregion
        #region 删除
        public static  void Delete(MODEL.Job job)
        {
            getJob.Delete(job);
        }
        #endregion
        public static DataSet getJobName(int id)
        {
            return getJob.getJobName(id);
        }
        public static MODEL.Job getJobModel(int id)
        {
            return getJob.getDeptModel(id);
        }
        public static bool checkJob(string name)
        {
          return getJob.checkJob(name);
        }
    }
}
```

BLLnotice.cs 中的具体代码如下：

```csharp
using System;
using System.Collections.Generic;
using System.Text;
```

```csharp
namespace BLL
{
    public class BLLnotice
    {
        public static readonly DAL.DALnotice getNotice=new DAL.DALnotice();
        ///<summary>
        ///公告
        ///</summary>
        #region 添加
        public static void Insert(MODEL.Notice notice)
        {
            getNotice.Insert(notice);
        }
        #endregion
        #region 更新
        public static  void Update(MODEL.Notice notice)
        {
            getNotice.Update(notice);
        }
        #endregion
        #region 删除
        public static  void Delete(MODEL.Notice notice)
        {
            getNotice.Delete(notice);
        }
        #endregion
        public static MODEL.Notice getNoteModel(int id)
        {
            return getNotice.getNoteModel(id);
        }
    }
}
```

BLLtask.cs 中的具体代码如下：

```csharp
using System;
using System.Collections.Generic;
using System.Text;

namespace BLL
{
    public class BLLtask
    {
        public static readonly DAL.DALtask getTask=new DAL.DALtask();
        ///<summary>
```

```
        ///工作任务
        ///</summary>
        #region 添加
        public static   void Insert(MODEL.Task task)
        {
            getTask.Insert(task);
        }
        #endregion
        #region 更新
        public static   void Update(MODEL.Task task)
        {
            getTask.Update(task);
        }
        #endregion
        #region 删除
        public static   void Delete(MODEL.Task task)
        {
            getTask.Delete(task);
         }
         #endregion
        #region  验证
        public static bool checkTask(string name)
        {
            return getTask.checkTask(name);
        }
         #endregion
    }
}
```

check.cs 中的具体代码如下:

```
using System;
using System.Collections.Generic;
using System.Text;
using DAL;
namespace BLL
{
    public class check
    {
        ///<summary>
        ///用户不存在或权限错误时跳转
        ///</summary>
        ///<param name="userName"></param>
```

```csharp
///<returns></returns>
public static string checkUserName(string userName)
{
    DALemployee employee=new DALemployee();
    return employee.loginUserLevel(userName);
}


///<summary>
///验证用户登录密码
///</summary>
///<param name="user">用户名</param>
///<param name="password">密码</param>

///<returns>"0"--用户名密码出错,"1"--普通员工,"2"--中层职位,"3"--高层职位</returns>
public static string checkUser(string user, string password)
{
    DALemployee employee=new DALemployee();
    string level=employee.loginUserLevel(user,password);

    return level;
}

///<summary>
///跟新在线状态
///</summary>
///<param name="state">"0"--下线,"1"--在线</param>
///<param name="name">姓名</param>
public static void updateOnLine(string state,string name)
{
    DAL.DALemployee employee=new DALemployee();
    employee.getOnline(state, name);
}
    }
}
```

7.8.6 系统实现——Web 层

本系统的登录界面如图 7.34 所示。

其中以任务计划模块为例讲解代码编写,该模块包括添加、删除任务,编辑任务状态等功能,具体界面如图 7.35 所示。

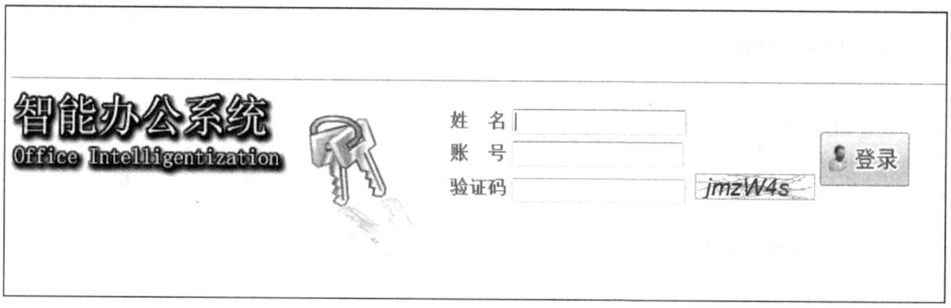

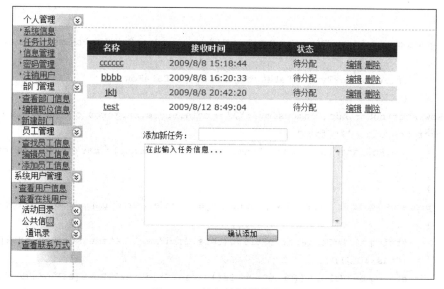

图 7.34　登录界面

图 7.35　任务计划模块界面

tasklist.ascx 页面的代码如下：

```
using System;
using System.Data;
using System.Configuration;
using System.Collections;
using System.Web;
using System.Web.Security;
using System.Web.UI;
using System.Web.UI.WebControls;
using System.Web.UI.WebControls.WebParts;
using System.Web.UI.HtmlControls;
public partial class UserControl_task : System.Web.UI.UserControl
{
    public int id= BLL.BLLemployee.getMobile(System.Web.HttpContext .Current.
Session["user"].ToString()).EmployeeID;
    protected void Page_Load(object sender, EventArgs e)
```

```csharp
    {
        if(!IsPostBack)
        {
            bind();
        }
    }
    public void bind()
    {
        SqlDataSource1.SelectParameters[0].DefaultValue=id.ToString();
        SqlDataSource1.Select(DataSourceSelectArguments.Empty);

    }
    protected void GridView1_RowDataBound(object sender, GridViewRowEventArgs e)
    {
        if(e.Row.RowType==DataControlRowType.DataRow)
        {
e.Row.Attributes.Add("onmouseover","c=this.style.backgroundColor;this.style.
backgroundColor='#ffff66'");
            e.Row.Attributes.Add("onmouseout", "this.style.backgroundColor=c");
        }
    }
    protected void GridView1_RowUpdating(object sender, GridViewUpdateEventArgs e)
    {
        string task=Convert.ToString(this.GridView1.DataKeys[e.RowIndex]
        ["taskID"]);
        string getValue= ((DropDownList)(this.GridView1.Rows[e.RowIndex].Cells
        [2].FindControl("DropDownList1"))).SelectedValue;
        SqlDataSource1.UpdateParameters[0].DefaultValue=getValue;
        SqlDataSource1.UpdateParameters[1].DefaultValue=task;
        SqlDataSource1.Update();
        bind();
    }
    protected void Button1_Click(object sender, EventArgs e)
    {
        if(this.CustomValidator1.IsValid==false)
        {
            msgBox.putMsg("名称有重复");
            return;
        }
        string taskName=this.TextBox1.Text;
        string taskInfo=this.TextArea1.Value;
        DateTime time=DateTime.Now;
        SqlDataSource1.InsertParameters[0].DefaultValue=taskName;
        SqlDataSource1.InsertParameters[1].DefaultValue=taskInfo;
```

```
        SqlDataSource1.InsertParameters[2].DefaultValue=time.ToString();
        SqlDataSource1.InsertParameters[3].DefaultValue=id.ToString();
        SqlDataSource1.Insert();
        bind();
        this.TextBox1.Text="";
        this.TextArea1.Value="";

    }
    protected void CustomValidator1_ServerValidate(object source,
    ServerValidateEventArgs args)
    {
        if(BLL.BLLtask.checkTask(args.Value)==false)
        {
            args.IsValid=false;
        }
        else
        {
            args.IsValid=true;
        }
    }
    protected void GridView1_RowDeleting(object sender, GridViewDeleteEventArgs e)
    {
        string task=Convert.ToString(this.GridView1.DataKeys[e.RowIndex]
        ["taskID"]);
        SqlDataSource1.DeleteParameters[0].DefaultValue=task;
        SqlDataSource1.Delete();
        bind();
    }
}
```

管理个人信息：包括填写个人信息、上传图面、填写就职协议，具体界面如图 7.36 所示。

personEdit.ascx 页面的代码如下：

```
using System;
using System.Data;
using System.Configuration;
using System.Collections;
using System.Web;
using System.Web.Security;
using System.Web.UI;
using System.Web.UI.WebControls;
using System.Web.UI.WebControls.WebParts;
using System.Web.UI.HtmlControls;
```

```csharp
public partial class UserControl_Edit : System.Web.UI.UserControl
{
    public MODEL.Employee employee;
    protected void Page_Load(object sender, EventArgs e)
    {
        if(!IsPostBack)
        {
            bindWorkDept();
            bindPage();
            bindWordJob();
        }
    }
    protected void Button1_Click(object sender, EventArgs e)
    {
            string name=Session["user"].ToString();
        employee=BLL.BLLemployee.getMobile(name);
        employee.Address=this.addTxt.Text;
        employee.EmployeeName=this.userName.Text;
        employee.JobID=Convert.ToInt32(this.jobList.SelectedValue);
        employee.DeptID=Convert.ToInt32(this.deptList.SelectedValue);
        employee.Email=this.email.Text;
        employee.Learn=this.learnList.SelectedItem.ToString();
        employee.Mobile=this.mobileTxt.Text;
        employee.NO=this.codeTxt.Text;
        employee.Photo=this.Image1.ImageUrl;
        employee.Sex=this.sexList.SelectedItem.ToString();
        employee.WorkState=this.stateList.SelectedItem.ToString();
        employee.UserLevel=this.levelList.SelectedValue;
        employee.Birthday=Convert.ToDateTime(this.birthTxt.Text);
        employee.AgreeMent=this.FCKeditor1.Value;
        BLL.BLLemployee.Update(employee);
        msgBox.putMsg("编辑成功","EditEmployee.aspx");
    }

    protected void Button2_Click(object sender, EventArgs e)
    {
        string fullName=this.fileUp.PostedFile.FileName.ToString();
        if(fullName=="")
        {
            msgBox.putMsg("请选择要上传的图片!");
            return;
        }
        if(this.userName.Text=="")
        {
```

```csharp
            msgBox.putMsg("您至少应输入姓名才能上传头像!");
            return;
        }
        string imgType=fullName.Substring(fullName.LastIndexOf("."));
        if(imgType==".jpg" || imgType==".bmp" || imgType==".gif")
        {
            ViewState["imgUrl"]=this.userName.Text+imgType;
            this.fileUp.PostedFile.SaveAs(Server.MapPath(" ~/photo/")+
            ViewState["imgUrl"]);
            this.Image1.ImageUrl="~/photo/"+ViewState["imgUrl"];
        }
        else
        {
            msgBox.putMsg("只能上传 jpg,gif,bmp 格式图片!");
        }
    }
    protected void deptList_SelectedIndexChanged(object sender, EventArgs e)
    {
        bindWordJob();
    }
    public void bindWordJob()
    {
        int deptID=Convert.ToInt32(this.deptList.SelectedValue);
        DataSet ds=BLL.BLLjob.getJobName(deptID);
        this.jobList.DataSource=ds.Tables[0].DefaultView;
        this.jobList.DataTextField="jobName";
        this.jobList.DataValueField="jobID";
        this.jobList.DataBind();
    }
    public void bindWorkDept()
    {
        DataSet ds=BLL.BLLdept.Dept();
        this.deptList.DataSource=ds.Tables[0].DefaultView;
        this.deptList.DataTextField="deptName";
        this.deptList.DataValueField="deptID";
        this.deptList.DataBind();
    }
    protected void bindPage()
    {
        if(Session["user"] !=null)
        {
            string name=Session["user"].ToString();
            employee=BLL.BLLemployee.getMobile(name);
            this.addTxt.Text=employee.Address;
```

```
                this.userName.Text=employee.EmployeeName;
                this.jobList.SelectedValue=Convert.ToString(employee.JobID);
                this.deptList.SelectedValue=Convert.ToString(employee.DeptID);
                this.email.Text=employee.Email;
                this.learnList.Text=employee.Learn;
                this.mobileTxt.Text=employee.Mobile;
                this.codeTxt.Text=employee.NO;
                this.Image1.ImageUrl=employee.Photo;
                this.sexList.Text=employee.Sex;
                this.stateList.Text=employee.WorkState;
                this.levelList.SelectedValue=employee.UserLevel;
                this.birthTxt.Text=employee.Birthday.ToShortDateString();
                this.FCKeditor1.Value=employee.AgreeMent;
            }
            else
            {
                Response.Redirect("BOSS.aspx");
            }
        }
        protected void Button3_Click(object sender, EventArgs e)
        {
            Response.Redirect("BOSS.aspx");
        }
    }
```

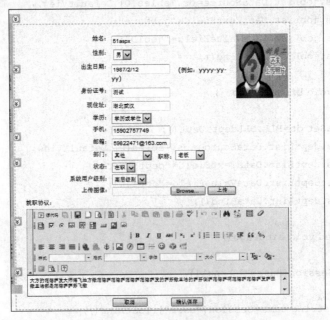

图 7.36 个人信息管理界面

另外,在 Web.config 中配置数据库连接语句,代码如下:

```xml
<?xml version="1.0"?>
<configuration>
    <system.web>
      <!--
        Set compilation debug="true" to insert debugging
        symbols into the compiled page. Because this
        affects performance, set this value to true only
        during development.
      -->
      <compilation debug="true" targetFramework="4.0">
        <assemblies>
          <add assembly="System.Web.Extensions, Version=4.0.0.0, Culture=neutral, PublicKeyToken=31bf3856ad364e35"/>
          <add assembly=" System. Design, Version = 4. 0. 0. 0, Culture = neutral, PublicKeyToken=B03F5F7F11D50A3A"/>
          <add assembly="System.Web.Extensions.Design, Version=4.0.0.0, Culture=neutral, PublicKeyToken=31BF3856AD364E35"/>
          <add assembly="System.Windows.Forms, Version=4.0.0.0, Culture=neutral, PublicKeyToken=B77A5C561934E089"/></assemblies>
      </compilation>
      <sessionState mode="InProc">
      </sessionState>
      <pages controlRenderingCompatibilityVersion=" 3. 5 " clientIDMode="AutoID"/></system.web>
    <system.web.extensions>
      <scripting>
        <webServices>
          <!-- Uncomment this line to customize maxJsonLength and add a custom converter-->
          <!--
    <jsonSerialization maxJsonLength="500">
      <converters>
        <add name="ConvertMe" type="Acme.SubAcme.ConvertMeTypeConverter"/>
      </converters>
    </jsonSerialization>
    -->
          <!--Uncomment this line to enable the authentication service. Include requireSSL="true" if appropriate.-->
          <!--
    <authenticationService enabled="true" requireSSL="true|false"/>
    -->
          <!--Uncomment these lines to enable the profile service. To allow
```

```
                    profile properties to be retrieved
                    and modified in ASP.NET AJAX applications, you need to add each
                    property name to the readAccessProperties and
                    writeAccessProperties attributes.-->
                        <!--
            <profileService enabled="true"
                        readAccessProperties="propertyname1,propertyname2"
                        writeAccessProperties="propertyname1,propertyname2" />
            -->
                </webServices>
                <!--
            <scriptResourceHandler enableCompression="true" enableCaching="true" />
            -->
                </scripting>
        </system.web.extensions>
        <connectionStrings>
        <add name="conn" connectionString="server=WIN-1TJ2F7NR33M\SQLEXPRESS;
        Integrated Security=SSPI;database=DatabaseOA"/>
        </connectionStrings>
        <appSettings>
            <add key="FCKeditor:BasePath" value="~/FCKeditor/"/>
            <add key="FCKeditor:UserFilesPath" value="~/image"/>
        </appSettings>
    </configuration>
```

第 8 章

大学毕业(论文)设计管理网站的研究与实现

毕业论文(设计)是高等教育中最后并且最重要的一个实践环节,它不是针对某一课题,而是要求综合运用本专业相关课程的理论和技术,设计出相应的解决问题的方案并加以实现。是学生走向工作岗位的前站,也是培养学生创新能力的重要途径之一。

根据需求对系统的功能进行了划分,总共 11 个主要功能,每个功能模块下都会有2～5个子功能模块,大部分的子功能模块都包含一个"审核"操作,该操作是管理员进行相关项目的审阅和意见填写的。功能模块划分如图 8.1 所示。

随着校园信息化建设的不断进展,很多高校已经采用信息化管理应用平台来进行毕业论文(设计)的管理工作。此举不仅能够简化操作流程、规范数据和文档的格式、方便数据的检索、提高管理效率,更能使学生和教师的沟通变得便捷有效。本项目是以大连大学毕业设计为需求分析对象,经过认真的调研了解,整个毕业设计的流程如图 8.2 所示。

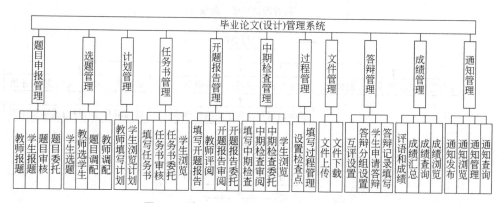

图 8.1 毕业论文(设计)管理系统功能划分

系统采用面向对象的设计方法,搭建 MVC 的开发框架,使用 Java 作为开发语言,MySQL 为数据库平台。

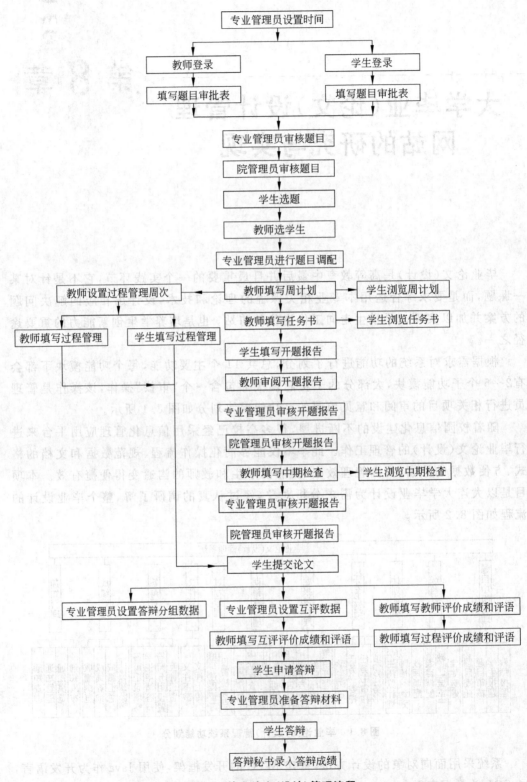

图 8.2 毕业论文(设计)管理流程

8.1 用例建模

系统的用户从使用角度来分析，主要有教师和学生两种用户；从管理的角度来分析，有专业级管理员、院级管理员、校级管理员和超级管理员五种角色，管理员可以由教师担任，并且管理的身份是可以叠加的，比如教师用户，既是普通的教师，也是专业管理员，也是院级管理员(见图 8.3)。

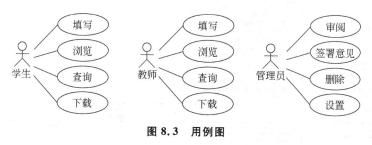

图 8.3 用例图

8.2 静态建模

类结构

根据 MVC 的设计思想，整个工程划分为三层，每个层都有相应的类支持，每层由不同的包标识。dao 包内定义的为数据库访问层的接口，dao.imp 包内定义的是数据库访问层的实现类，action 包内定义的控制层的类，domain 包内定义的为数据库模型层的类；工程下包的定义如图 8.4 所示，baset 包内定义的是公共类，interceptor 包内定义的是拦截器类，servlet 包内定义的类是用来在系统启动的时候对系统进行参数初始化的，test 包内定义的是用来进行单元测试的类，util 包内定义的是工具类。

图 8.4 包的定义

base 包内定义的是公共类,主要用来被继承的,类图如图 8.5 所示。

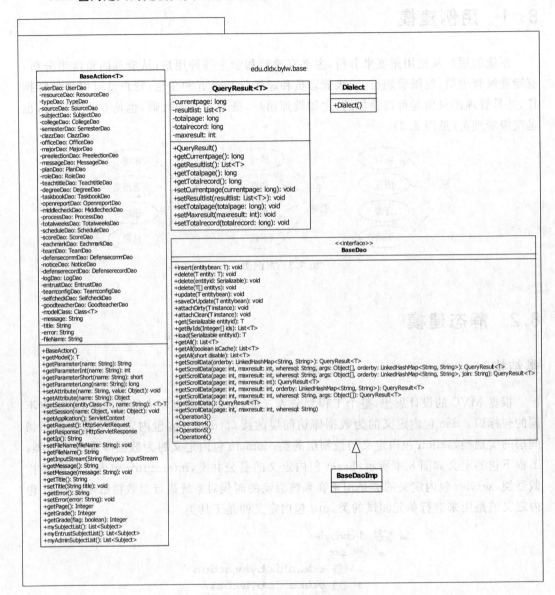

图 8.5　base 包所定义类的类图

以 UserDao 为例,图 8.6 描述了 dao 类的结构关系。

以 User、role 类为例,图 8.7 描述了 domain 类的结构关系,由于篇幅关系,domain 类的类图中指描述的属性,省略了每个属性都应该有的 getter 和 setter 方法。

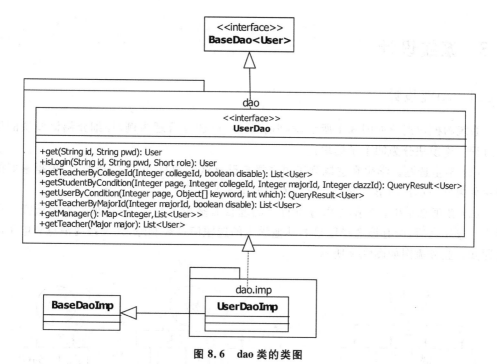

图 8.6　dao 类的类图

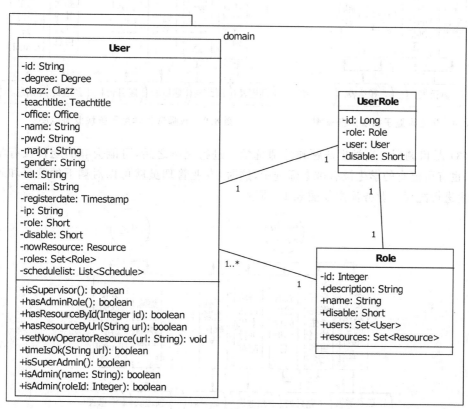

图 8.7　domain 类的类图

8.3 系统设计

8.3.1 功能设计

系统的模块划分如图 8.1 所示,本小节开始,以选题管理为例,详细介绍该模块的设计过程。该模块分为四个子模块:

(1)学生选题。学生在选题列表中选择自己喜欢的题目,可以选择两个,根据喜好的程序分为第一选择和第二选择。业务流图如图 8.8 所示。

(2)教师选学生。教师可以查看自己的题目有多少人报名,并在报名人中选择一位学生。选择之后,所有该选择题目的其他学生的选题信息被删除,并发站内消息和电子邮件通知。业务流图如图 8.9 所示。

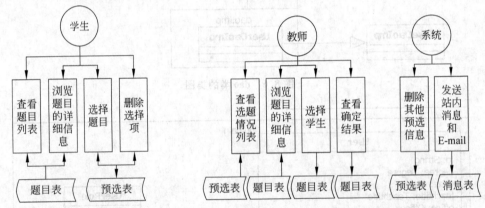

图 8.8 学生选题子模块业务流图　　图 8.9 教师选择学生子模块业务流图

(3)题目调配。当学生选题和教师选学生进行完毕之后,可能会有的题目没有学生选择,也有可能有的学生没有被老师选中,这时,专业管理员就可以对剩下的题目和剩下的学生进行配对。业务流图如图 8.10 所示。

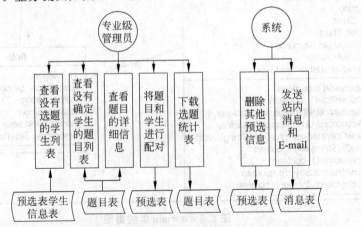

图 8.10 题目调配子模块业务流图

（4）调配指导教师。这个模块是针对现在校外实习的学生而设计的，这部分学生在企业实习或培训，可以结合自己正在实习或者培训的内容，自己拟定一个论文题目。但这样的题目就没有校内的指导教师，因此需要给这些题目分配一个校内的指导教师。业务流图如图 8.11 所示。

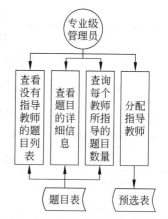

图 8.11　调配指导教师子模块业务流图

8.3.2　数据库设计

本系统总共建立了 37 张数据表，具体如表 8.1 所示。

表 8.1　数据表清单

序号	表名		序号	表名	
1	clazz	班级信息表	20	resource	资源表
2	college	学院信息表	21	role	角色表
3	comment	评语表	22	roles_resource	角色资源对照表
4	defensecomm	答辩组组员信息表	23	schedult	日程安排表
5	defenserecord	答辩记录表	24	school	学校信息表
6	degree	学历信息表	25	score	成绩表
7	eachmark	互评分配表	26	selfcheck	自查表
8	entrust	委托表	27	source	题目来源信息表
9	filedownload	下载文件表	28	status	题目状态表
10	goodteacher	优秀指导教师信息表	29	subject	题目信息表
11	log	日志表	30	taskbook	任务书表
12	message	站内消息表	31	teachtitle	职称信息表
13	middlecheck	中期检查表	32	team	答辩分组表
14	notice	通知表	33	teamconfig	答辩分组配置信息表
15	office	办公室信息表	34	totalweeks	周期表
16	openreport	开题报告表	35	type	题目类型表
17	plan	计划表	36	user	用户表
18	prelection	选题信息表	37	users_roles	用户角色对照表
19	process	进度管理表			

选题模块主要涉及四张表，分别为 user、subject、prelection、message 表。表结构如表 8.2～表 8.5 所示。

表 8.2 user 数据表

序号	字段名	数据类型	是否为主键	注释
1	id	varchar(8)	是	用户名
2	name	varchar(20)	否	姓名
3	pwd	varchar(50)	否	密码
4	teachtitleId	int(10) unsigned	否	职称 ID
5	officeId	int(10) unsigned	否	办公室 ID
6	major	varchar(20)	否	专业
7	degreeId	int(10) unsigned	否	学位
8	gender	varchar(2)	否	性别
9	tel	varchar(30)	否	电话
10	email	varchar(50)	否	邮箱
11	registerdate	timestamp	否	注册日期
12	ip	varchar(20)	否	最近一次登录 IP
13	clazzId	int(10) unsigned	否	所属班级
14	role	tinyint(3)	否	0-教师 1-学生
15	disable	tinyint(3) unsigned	否	0-无效 1-有效

表 8.3 subject 数据表

序号	字段名	数据类型	是否为主键	注释
1	id	varchar(15)	是	题目 id
2	tea1Id	varchar(8)	否	第一指导教师
3	studentId	varchar(8)	否	学生
4	tea2Id	varchar(8)	否	第二指导教师
5	name	varchar(50)	否	题目名
6	typeId	int(10)	否	题目类型
7	first	tinyint(1)	否	0-不是第一次指导,1-第一次指导
8	same	tinyint(1)	否	0-不雷同,1-雷同
9	personcount	tinyint(3)	否	拟需学生人数
10	grade	int(5)	否	届
11	sourceId	int(10) unsigned	否	题目来源
12	summary	varchar(255)	否	选题内容提要
13	feasibility	varchar(255)	否	可行性

续表

序号	字段名	数据类型	是否为主键	注释
14	necessary	varchar(255)	否	必要性
15	majorId	int(10)	否	题目给哪个专业学生选的
16	date	timestamp	否	提交日期
17	comment1	varchar(255)	否	教研室主任意见
18	person1	varchar(8)	否	教研室主任意见签署人
19	date1	timestamp	否	教研室主任审核意见签署时间
20	comment2	varchar(255)	否	学院指导委员会意见
21	person2	varchar(8)	否	学院指导委员会意见签署人
22	date2	timestamp	否	学院指导委员会意见签署日期
23	ordernum	smallint(6)	否	查档编号
24	disable	tinyint(3)	否	0-有效,1-无效
25	status	tinyint(4)	否	状态
26	othermajor	varchar(200)	否	其他可选专业

表 8.4 prelection 数据表

序号	字段名	数据类型	是否主键	注释
1	id	bigint(20) unsigned	是	
2	subjectId	varchar(15)	否	题目 ID
3	studentId	varchar(8)	否	学生 ID
4	choice	tinyint(3)	否	第几选择(每个学生允许有两个选择)
5	selectdate	timestamp	否	操作日期
6	disable	tinyint(3)	否	0-有效 1-无效

表 8.5 message 数据表

序号	字段名	数据类型	是否主键	注释
1	id	bigint(20)	是	ID
2	title	varchar(60)	否	标题
3	content	text	否	内容
4	sender	varchar(8)	否	发送人 ID

续表

序号	字段名	数据类型	是否主键	注 释
5	receiver	varchar(8)	否	接收人 ID
6	sendtime	timestamp	否	发送时间
7	senderdisable	tinyint(4)	否	发送人删除标志
8	receiverdisable	tinyint(4)	否	接收人删除标志
9	hasread	tinyint(4)	否	阅读标志 1-未读 0-已读
10	parentId	bigint(20)	否	父 ID
11	level	tinyint(3)	否	0-一般消息 1-专业级消息 2-院级消息 3-校级消息

表之间的关系如图 8.12 所示。

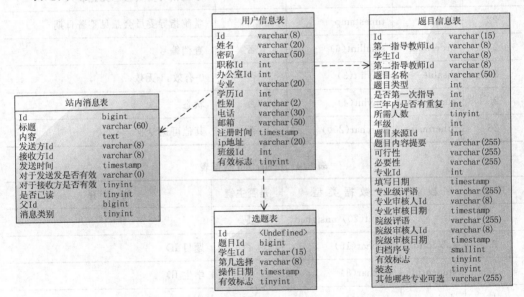

图 8.12　subject、prelection、user、message 表之间的关系

8.4　动态建模

(1) 学生选题模块，如图 8.13 所示。

(2) 教师确认学生，如图 8.14 所示。

(3) 题目调配，如图 8.15 所示。

(4) 调配教师，如图 8.16 所示。

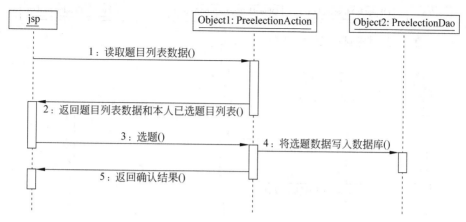

图 8.13 学生选题模块顺序图

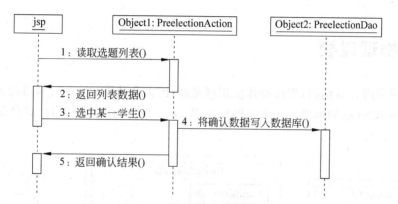

图 8.14 教师确认学生顺序图

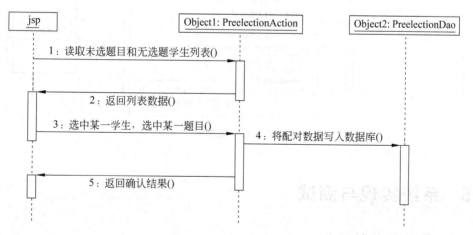

图 8.15 题目调配顺序图

图 8.16　调配教师顺序图

8.5　物理建模

该系统采用 B/S 结构，界面操作使用浏览器，因大量使用 CSS 和 JS，因此浏览器要求使用符合 W3C 标准的浏览器。服务器端使用 Tomcat 作为服务器软件，具体的部署关系如图 8.17 所示。

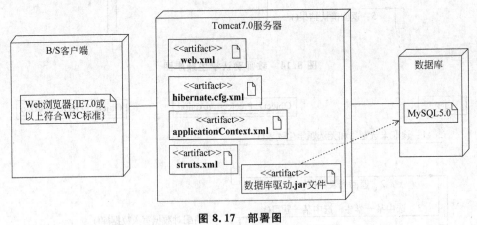

图 8.17　部署图

8.6　系统实现与测试

8.6.1　公共部分的设计

1. domain 包内的 User 类

```
package edu.dldx.bylw.domain;
```

```java
import java.sql.Timestamp;
import java.util.ArrayList;
import java.util.HashSet;
import java.util.List;
import java.util.Set;
import javax.persistence.Entity;
import com.opensymphony.xwork2.ActionContext;
import edu.dldx.bylw.util.Sys;
import edu.dldx.bylw.util.Util;
@SuppressWarnings("serial")
@Entity
public class User implements java.io.Serializable,Comparable<User>{
    private String id;                              //工号
    private Degree degree;                          //教师的学位
    private Clazz clazz;                            //学生的班级
    private Teachtitle teachtitle;                  //职称
    private Office office;                          //教师的教研室
    private String name;                            //姓名
    private String pwd;                             //密码
    private String major;                           //教师的专业
    private String gender;                          //性别
    private String tel;                             //电话
    private String email;                           //邮件
    private Timestamp registerdate;                 //注册日期
    private String ip;                              //IP地址
    private Short role;                             //身份角色 0-教师 1-学生
    private Short disable;                          //有效无效
    private Resource nowResource;                   //正在操作的栏目
    private Set<Role> roles=new HashSet<Role>(0);   //用户操作角色
    private List<Schedule> schedulelist=new ArrayList<Schedule>();
    //判断是否是督导
    public boolean isSupervisor(){
        if(isSuperAdmin())
            return true;
        for(Role r:roles){
            String roleId=r.getName();
            if(roleId.equals(Role.SUPERVISOR))
                return true;
        }
        return false;
    }
    public boolean hasAdminRole(){
        if(isSuperAdmin())
            return true;
```

```java
            for(Role r:roles){
                String roleId=r.getName();
                if(roleId.equals(Role.SCHOOL)||roleId.equals(Role.COLLEGE)||
                roleId.equals(Role.MAJOR))
                    return true;
            }
            return false;
        }
        public boolean hasResourceById(Integer id){      //是否是有某个资源的权限
            if(isSuperAdmin())
                return true;
            for(Role r:roles){
                for(Resource resource:r.getResources()){
                    if(id==resource.getId()){
                        return true;
                    }
                }
            }
            return false;
        }
        @SuppressWarnings("unchecked")
        public boolean hasResourceByUrl(String url){     //是否是有某个url的权限
            this.setNowResource(null);
            if(isSuperAdmin()){
                this.setNowOperatorResource(url);
                return true;
            }
            //如果是公共功能(登录用户后都可以使用的功能,不需要控制),就返回true
            List<String> allPrivilegeUrls=(List<String>)ActionContext.getContext().
            getApplication().get("allUrls");
            if(!allPrivilegeUrls.contains(url)){
                //不在allPrivilegeUrls的地址都算不受控制的地址
                this.setNowOperatorResource(url);
                return true;
            }
            for(Role r:roles){
                for(Resource resource:r.getResources()){
                    if(resource.getParent().getId()==0)
                        continue;
                    if(url.equals(resource.getAddress())){
                        this.setNowResource(resource);
                        return true;
                    }
                }
```

```java
        }
        return false;
}
public void setNowOperatorResource(String url){
    for(Role r:roles){
        for(Resource resource:r.getResources()){
            if(resource.getParent().getId()==0)
                continue;
            if(url.equals(resource.getAddress())){
                this.setNowResource(resource);
                return;
            }
        }
    }
}
public boolean timeIsOk(String url){
    if(url.indexOf("login")>=0||url.indexOf("index")>=0)
        return true;
    if(this.schedulelist.size()<=0)
        return false;
    else
        for(Schedule s:this.schedulelist){
            if(url.equals(s.getResource().getAddress())){
                long now=System.currentTimeMillis();
                if(now< s.getBegindate().getTime()||now> s.getEnddate().getTime())
                    return false;
            }
        }
    return true;
}
public boolean isSuperAdmin(){
    return id!=null&&id.equals(Sys.getSuperAdmin());
}
public boolean isAdmin(String name){
    for(Role r:roles){
        if(name.equals(r.getName()))
            return true;
    }
    return false;
}
public boolean isAdmin(Integer roleId){
    for(Role r:roles){
        if(r.getId()>roleId)
```

```java
            return true;
    }
    return false;
}

/**省略 getter 和 setter 的代码,请读者编程的时候自行加上 */

@Override
public boolean equals(Object obj){
    if(this==obj)
        return true;
    if(obj==null)
        return false;
    if(getClass()!=obj.getClass())
        return false;
    User other=(User)obj;
    if(id==null){
        if(other.id!=null)
            return false;
    }else if(!id.equals(other.id))
        return false;
    return true;
}
@Override
public int compareTo(User o){
    return this.getId().compareTo(o.getId());
}
public boolean messageCompleted(){
    if(Util.isEmpty(this.tel))
        return false;
    if(Util.isEmpty(this.gender))
        return false;
    if(Util.isEmpty(this.email))
        return false;
    if(this.role==User.TEACHER){                    //教师
        if(this.teachtitle==null)
            return false;
        if(this.office==null||this.office.getId()==null)
            return false;
        if(this.degree==null)
            return false;
    }else{                                          //学生
        if(this.clazz==null)
            return false;
```

            }
            return true;
      }
}

2. domain 包内的 Subject 类

```
package edu.dldx.bylw.domain;
import java.sql.Timestamp;
import java.util.HashSet;
import java.util.Set;
import javax.persistence.Entity;
@SuppressWarnings("serial")
@Entity
public class Subject implements java.io.Serializable{
      private String id;
      private Source source;
      private User person1;
      private User tea1;
      private User person2;
      private User tea2;
      private Type type;
      private String name;
      private Short first;              //0代表不是第一次指导,1是第一次指导
      private Short same;               //是否有雷同题目 0代表不雷同,1代表雷同
      private Short personcount;        //拟需学生人数
      private String summary;
      private String feasibility;
      private String necessary;
      private User student;             //学生
      private Major major;              //题目是分配给哪个专业学生做的
      private Integer grade;            //届
      private Timestamp date;           //录入时间
      private String comment1;
      private Timestamp date1;          //系审核人签字时间
      private String comment2;
      private Timestamp date2;          //院审核人签字时间
      private Short ordernum;
      private Short disable;
      private String othermajor;        //本院其他专业是否可选
      private Integer status;           //状态
      private Openreport openreport;    //开题报告
      private Middlecheck middlecheck;  //中期检查
      private Taskbook taskbook;        //任务书
      private Set<Plan>plans=new HashSet<Plan>(0);
```

/**省略 getter 和 setter 的代码,请读者编程的时候自行加上 */
}

3. domain 包下的 Preelection 类

```java
package edu.dldx.bylw.domain;
import java.sql.Timestamp;
import javax.persistence.Entity;
@SuppressWarnings("serial")
@Entity
public class Preelection implements java.io.Serializable{
    private Long id;
    private Subject subject;
    private User user;
    private Short choice;
    private Timestamp selectdate;
    private Short disable;

    /**省略 getter 和 setter 的代码,请读者编程的时候自行加上 */
}
```

4. domain 包下的 message 类

```java
package edu.dldx.bylw.domain;
import java.sql.Timestamp;
import java.util.HashSet;
import java.util.Set;
import javax.persistence.Entity;
/**
 * 站内消息
 * @author qsm
 */
@SuppressWarnings("serial")
@Entity
public class Message implements java.io.Serializable{
    private Long id;
    private User receiver;
    private User sender;
    private String title;
    private String content;
    private Timestamp sendtime;
    private Short senderdisable;
    private Short receiverdisable;
    private Short hasread;
```

```
    private Message parent;
    private Integer level;
    private Set<Message> children=new HashSet<Message>(0);

    /**省略 getter 和 setter 的代码,请读者编程的时候自行加上 */
}
```

5. action 包下定义的 PreelectionAction 类

```
package edu.dldx.bylw.action;
import java.sql.Timestamp;
import java.util.List;
import java.util.Map;
import org.springframework.context.annotation.Scope;
import org.springframework.stereotype.Controller;
import com.opensymphony.xwork2.ActionContext;
import edu.dldx.bylw.base.BaseAction;
import edu.dldx.bylw.base.QueryResult;
import edu.dldx.bylw.domain.College;
import edu.dldx.bylw.domain.Major;
import edu.dldx.bylw.domain.Preelection;
import edu.dldx.bylw.domain.Role;
import edu.dldx.bylw.domain.Status;
import edu.dldx.bylw.domain.Subject;
import edu.dldx.bylw.domain.User;
import edu.dldx.bylw.util.Sys;
import edu.dldx.bylw.util.Util;
@SuppressWarnings("serial")
@Controller("preelectionAction")
@Scope("prototype")
public class PreelectionAction extends BaseAction<Preelection>{
    //题目调配列表
    public String adminList(){
        Integer majorId=this.getParameterInt("majorId");
        Integer collegeId=this.getParameterInt("collegeId");
        User user=getSession(User.class,"user");
        if(collegeId==-1&&majorId==-1){
            collegeId=user.getOffice().getMajor().getCollege().getId();
            majorId=user.getOffice().getMajor().getId();
        }
        //准备管理员的查询数据
        if(user.isAdmin(Role.SCHOOL)){
            //学院列表
        }else if(user.isAdmin(Role.COLLEGE)){
            College college = collegeDao.get(user.getOffice().getMajor().
```

```java
        getCollege().getId());
    this.setAttribute("majorlist",college.getMajors());
}
this.setAttribute("collegeId",collegeId);        //当前列表数据是那个院的
this.setAttribute("majorId",majorId);            //当前列表数据是那个专业的
int which=super.getParameterInt("which");        //0-调配老师 1-调整学生选题
which=which<0?1:which;
Integer grade=super.getGrade();
List<Subject>subjectlist=null;
String keyword=super.getParameter("keyword");
switch(which){
case 0:                                          //0-调配老师
    subjectlist = preelectionDao. getSubjectlisthasNoTeacherByMajorId
    (majorId,grade,keyword);                     //没有指导教师的题目
    Map<String,Long>map=subjectDao.getSubjectCountByMajorId(majorId,
    grade);                                      //教师工号-数量
    List<User>teacherlist=userDao.getTeacherByMajorId(majorId,true);
    this.setAttribute("teacherlist",teacherlist);
    this.setAttribute("map",map);
    break;
case 1:                                          //1-调整学生选题
    //先查询本专业当前没有被选中的题目
    subjectlist = preelectionDao. getSubjectlistNotSelectedByMajorId
    (majorId,grade);                             //没有被选中的题目
    //再查询本专业还没有题目的学生
    List< User > studentlist = preelectionDao. getStudentlistNoSubjectByMajorId
    (majorId,grade);
    this.setAttribute("studentlist",studentlist); //没有题目的学生
}
this.setAttribute("keyword",keyword==null?"":keyword);
this.setAttribute("which",which);                //0-调配老师 1-调整学生选题
this.setAttribute("subjectlist",subjectlist);
return "adminList";
}
/** 调配学生选题 */
public String admin(){
    int which=super.getParameterInt("which");
    String subjectId=super.getParameter("subjectId");
    Subject subject=null;
    switch(which){
    case 0:                                      //调整教师
        subject=subjectDao.get(subjectId);
        subject.setTea1(model.getUser());
        subjectDao.update(subject);
```

```java
            super.setMessage("数据保存成功!");
            break;
        case 1:                                             //给学生分配题目
            String studentId=super.getParameter("studentId");
            subject=subjectDao.getByStudentId(studentId,Sys.getGrade());
            if(subject!=null){                              //该学生已经有题目了
                super.setMessage("2");
                return "ajaxjson";
            }
            subject=subjectDao.get(subjectId);
            if(subject.getStudent()==null){
                User student=new User();
                student.setId(studentId);
                subject.setStudent(student);
                subject.setStatus(Status.SELECTED);
                subjectDao.update(subject);
                super.setMessage("1");                      //成功
            }else{
                super.setMessage("3");                      //该题目已经被有学生了
            }
            User user=super.getSession(User.class,"user");
            logDao.debug("管理员将题目:"+subjectId+"分配给学生:"+studentId,
                user.getId(),super.getIp());
            break;
    }
    return "ajaxjson";
}
/** 选题列表 */
public String list(){
    User user=super.getSession(User.class,"user");
    switch(user.getRole()){
    case User.STUDENT:
        return this.selectlist();
    case User.TEACHER:
        return this.confirmlist();
    }
    return "list";
}
/** 添加页面 */
public String addView(){
    return "addUI";
}
/** 添加 */
public String add(){
```

```java
        User user=super.getSession(User.class,"user");
        Long times = preelectionDao.getTimes (user.getId (),model.getChoice (),
        Sys.getGrade());
        if(times>=Sys.getPreelectionTimes()){
            super.setMessage("修改次数已经达到上限!");
            return "ajaxjson";
        }
        preelectionDao.disable(user.getId(),model.getChoice());
                                                        //先将以前选择的取消
        Preelection p=new Preelection();
        p.setSubject(model.getSubject());
        p.setSelectdate(new Timestamp(System.currentTimeMillis()));
        p.setDisable((short)1);
        p.setUser(user);
        p.setChoice(model.getChoice());
        preelectionDao.insert(p);
        super.setMessage("");
        return "ajaxjson";
    }
    /** 学生选题列表 */
    private String selectlist(){
        Integer grade=Sys.getGrade();                   //届
        User user=super.getSession(User.class,"user");
        Subject subject=subjectDao.getByStudentId(user.getId(),Sys.getGrade());
        if(subject!=null){
            super.setMessage("您的题目已经确定!");
            subject.setSummary(Util.indent(subject.getSummary()));
            subject.setFeasibility(Util.indent(subject.getFeasibility()));
            subject.setNecessary(Util.indent(subject.getNecessary()));
            ActionContext.getContext().getValueStack().push(subject);
            return "detail";
        }
        Integer[] times=preelectionDao.getTimes(user.getId(),super.getGrade());
        if ( times [ 0 ] > = Sys. getPreelectionTimes ( ) &&times [ 1 ] > = Sys.
        getPreelectionTimes()){
            super.setMessage("您的选题修改次数已达到上限,无法再次修改,请等待老师
            确认后的结果。");
            super.setAttribute("result",null);          //题目列表
        }else{
            String keyword=super.getParameter("keyword");
            QueryResult<Subject> qr=null;
            Major major = user. getClazz (). getMajor ();  //clazzDao. get (user.
            getClazz().getId()).getMajor().getId();
            if(!Util.isEmpty(keyword)&&!"按教师姓名或题目名称模糊查询".equals
```

```java
            (keyword)){
                qr=subjectDao.getSubjectListForSelect(super.getPage(),grade,
                major,keyword);
                super.setAttribute("keyword",keyword);      //题目列表
            }else
                qr=subjectDao.getSubjectListForSelect(super.getPage(),grade,
                major);
            Map<String,Integer> nummap = subjectDao.getPreelectionNum (qr.
            getResultlist());
            super.setAttribute("result",qr);              //题目列表
            super.setAttribute("nummap",nummap);          //每个题目的选择数量
        }
        List<Preelection> preelectionlist = preelectionDao.getPreelectionlist
        (user.getId());
        super.setAttribute("preelectionlist",preelectionlist);    //已选题列表
        super.setAttribute("times",times);                //选题的修改次数
        return "selectlist";
    }
    /**确认学生列表 */
    public String confirmlist(){
        User user=super.getSession(User.class,"user");
        QueryResult<Subject> qr=subjectDao.getMySubjectList(super.getPage(),
        super.getGrade(),user.getId());
        super.setAttribute("result",qr);
        Map<String,List<String[]>>map1=preelectionDao.getPreelectionStudent
        (qr.getResultlist(),(short)1);
            Map<String,List<String[]>> map2 = preelectionDao.
        getPreelectionStudent(qr.getResultlist(),(short)2);
        super.setAttribute("map1",map1);              //第一选择情况
        super.setAttribute("map2",map2);              //第二选择情况
        return "confirmlist";
    }
}
```

6. dao 包下定义的 PreelectionDao 接口

```java
package edu.dldx.bylw.dao;
import java.util.List;
import java.util.Map;
import edu.dldx.bylw.base.BaseDao;
import edu.dldx.bylw.domain.Preelection;
import edu.dldx.bylw.domain.Subject;
import edu.dldx.bylw.domain.User;
public interface PreelectionDao extends BaseDao<Preelection>{
    /**获得某学生的预选情况 */
```

```java
    public List<Preelection>getPreelectionlist(String studentId);
    /**获得某教师的预选情况*/
    public List < Preelection > getPreelectionlist (String studentId, Integer grade);
    /**获得某个学生的第n个志愿选题情况*/
    public Preelection get(String studentId,Short n);
    /**将某个学生的第n个志愿选题情况的disable标志设置为0*/
    public void disable(String studentId,Short n);
    /**将某个题目的预选情况的disable标志设置为0*/
    public void disable(String subjectId);
    /**获得某个学生的第n个志愿选题的次数*/
    public Long getTimes(String studentId,Short n,Integer grade);
    /**获得某个学生志愿选题的次数*/
    public Integer[] getTimes(String studentId,Integer grade);
    /**获得题目的预选学生列表*/
    public Map< String, List< String[]>> getPreelectionStudent (List< Subject> subjectlist,Short choice);
    /**查询本专业当前没有被选中的题目*/
    public List < Subject > getSubjectlistNotSelectedByMajorId (Integer majorId, Integer grade);
    /**查询本专业还没有题目的学生*/
    public List<User>getStudentlistNoSubjectByMajorId(Integer majorId,Integer grade);
    /**获得某专业还没有指定教师的题目*/
    public List<Subject>getSubjectlisthasNoTeacherByMajorId(Integer majorId, Integer grade,String keyword);
}
```

7. dao.imp 包下定义的 PreelectionDaoImp 类

```java
package edu.dldx.bylw.dao.imp;
import java.util.ArrayList;
import java.util.HashMap;
import java.util.List;
import java.util.Map;
import org.hibernate.type.IntegerType;
import org.hibernate.type.StringType;
import org.springframework.stereotype.Repository;
import org.springframework.transaction.annotation.Transactional;
import edu.dldx.bylw.base.BaseDaoImp;
import edu.dldx.bylw.dao.PreelectionDao;
import edu.dldx.bylw.domain.Preelection;
import edu.dldx.bylw.domain.Subject;
import edu.dldx.bylw.domain.User;
import edu.dldx.bylw.util.Util;
```

```java
@SuppressWarnings("unchecked")
@Repository("preelectionDao")
@Transactional
public class PreelectionDaoImp extends BaseDaoImp<Preelection> implements PreelectionDao{
    /**获得某学生的预选情况*/
    public List<Preelection>getPreelectionlist(String studentId){
        List<Preelection>list=getSession().createQuery(//
            "From Preelection o where o.user.id=? and o.disable=1 order by o.choice")//
            .setParameter(0,studentId)//
            .list();
        if(list!=null&&list.size()<=2){
            for(int i=0;i<list.size();i++){
                Preelection p=list.get(i);
                if(i==p.getChoice()-1)
                    continue;
                else{
                    Preelection temp=new Preelection();
                    temp.setChoice((short)(i+1));
                    list.add(i,temp);
                }
            }
        }
        return list;
    }
    /**获得某教师的预选情况*/
    public List<Preelection> getPreelectionlist(String studentId, Integer grade){
        return null;
    }
    /**获得某个学生的第n个志愿选题情况*/
    public Preelection get(String studentId,Short n){
        return(Preelection)getSession().createQuery(//
            "From Preelection o where o.user.id=? and o.disable=1 and o.choice=?")//
            .setParameter(0,studentId).setParameter(1,n)//
            .uniqueResult();
    }
    /**将某个学生的第n个志愿选题情况的disable标志设置为0*/
    public void disable(String studentId,Short n){
        getSession().createQuery(//
            "update Preelection o set o.disable=0 where o.user.id=?"+(n==-1?"":" and o.choice="+n))//
```

```java
            .setParameter(0,studentId)//
            .executeUpdate();
}
/**将某个题目的预选情况的 disable 标志设置为 0 */
public void disable(String subjectId){
    getSession().createQuery(//
        "update Preelection o set o.disable=0 where o.subject.id=?")//
        .setParameter(0,subjectId)//
        .executeUpdate();
}
/**获得某个学生的第 n 个志愿选题的次数 */
public Long getTimes(String studentId,Short n,Integer grade){
    return(Long)getSession().createQuery(//
        "select count (id) from Preelection o where o.user.id=? and o.choice=? and o.subject.grade=?")//
        .setParameter(0,studentId).setParameter(1,n).setParameter(2,grade)//
        .uniqueResult();
}
/**获得某个学生志愿选题的次数 */
public Integer[] getTimes(String studentId,Integer grade){
    List<Object[]>list=getSession().createSQLQuery(//
        "select choice as c, count (id) as num from preelection o where studentId=? and left(subjectId,4)=? group by choice")//
        .addScalar("c", new IntegerType()).addScalar("num", new IntegerType())//
        .setParameter(0,studentId)//
        .setParameter(1,grade)//
        .list();
    Integer[] result={0,0};
    for(int i=0;i<list.size();i++){
        Object[] arr=list.get(i);
        result[(Integer)arr[0]-1]=(Integer)arr[1];
    }
    return result;
}
/**获得题目的预选学生列表 */
public Map<String,List<String[]>> getPreelectionStudent(List<Subject> subjectlist,Short choice){
    if(subjectlist==null||subjectlist.size()<=0)
        return null;
    Map<String,List<String[]>>map=new HashMap<String,List<String[]>>();
    StringBuffer sql=new StringBuffer("select p.subjectId as id,student.id as sid,student.name as sname,c.name as cn ");
```

```java
sql.append("from preelection p left join user student on p.studentId=student.id ")//
        .append("left join clazz c on c.id=student.clazzId ")//
        .append("where subjectId IN('");
StringBuffer tempsql=new StringBuffer();
for(Subject subject:subjectlist){
    tempsql.append(subject.getId()).append("','");
}
sql.append(tempsql.length()<=0?"'":tempsql.substring(0,tempsql.length()-2));
sql.append(") and p.choice=").append(choice).append(" and p.disable=1").append(" order by p.subjectId desc");
List<Object[]> list= super.getSession().createSQLQuery(sql.toString())//
        .addScalar("id", new StringType()).addScalar("sid", new StringType())//
        .addScalar("sname", new StringType()).addScalar("cn", new StringType())//
        .list();
if(list==null||list.size()<=0)
    return null;
String subjectId=(String)list.get(0)[0];
List<String[]> userlist=new ArrayList<String[]>();
for(int i=0;i<list.size();i++){
    Object[] obj=list.get(i);
    String[] userarr=new String[3];
    if(!subjectId.equals((String)obj[0])){
        map.put(subjectId,userlist);
        subjectId=(String)obj[0];
        userlist=new ArrayList<String[]>();
    }
    for(int j=0;j<userarr.length;j++)
        userarr[j]=(String)obj[j+1];
    userlist.add(userarr);
}
map.put(subjectId,userlist);
return map;
}
/**查询本专业当前没有被选中的题目*/
public List<Subject> getSubjectlistNotSelectedByMajorId(Integer majorId,Integer grade){
    return getSession().createQuery(//
            "From Subject s where s.major.id=? and s.student=null and s.grade=?")//
```

```java
            .setParameter(0,majorId)//
            .setParameter(1,grade)//
            .list();
}
/**查询本专业还没有题目的学生 */
public List<User>getStudentlistNoSubjectByMajorId(Integer majorId,Integer grade){
    List<String>userlist=getSession().createQuery(//
        "Select s.student.id From Subject s where s.student is not null and s.student.clazz.major.id=? and s.student.clazz.grade=?")//
        .setParameter(0,majorId)//
        .setParameter(1,grade-4)//
        .list();
    if(userlist!=null&&userlist.size()>0){
        List<User>list=getSession().createQuery(//
            "From User u where u.role=1 and u.clazz.major.id=? and u.clazz.grade=? and u.id not in(:ids)and u.disable=1 ")//
            .setParameter(0,majorId)//
            .setParameter(1,grade-4)//
            .setParameterList("ids",userlist.toArray())//
            .list();
        return list;
    }
    return null;
}
/** 获得某专业还没有指定教师的题目 */
public List<Subject> getSubjectlisthasNoTeacherByMajorId(Integer majorId,Integer grade,String keyword){
    StringBuffer hql=new StringBuffer("From Subject s where s.major.id=? and s.tea1=null and s.grade=? and s.student!=null and s.disable=1");
    if(!Util.isEmpty(keyword))
        hql.append(" and(s.student.id='").append(keyword).append("' or s.student.name like '%").append(keyword).append("%')");
    return getSession().createQuery(//
        hql.toString())//
        .setParameter(0,majorId)//
        .setParameter(1,grade)//
        .list();
}
```

8. Preelection 类的 hibernate 配置文件

```xml
<?xml version="1.0"?>
<!DOCTYPE hibernate-mapping PUBLIC
```

```
"-//Hibernate/Hibernate Mapping DTD 3.0//EN"
"http://www.hibernate.org/dtd/hibernate-mapping-3.0.dtd">
<hibernate-mapping package="edu.dldx.bylw.domain">
    <class name="Preelection" table="preelection">
        <cache usage="read-only"/>
        <id name="id">
            <generator class="native" />
        </id>
        <property name="choice" />
        <property name="selectdate" />
        <property name="disable" />
        <!--subject 属性，this 与 subject 的多对一关系-->
        <many-to-one name="subject" class="Subject" column="subjectId" />
        <!--user 属性，this 与 user 的多对一关系-->
        <many-to-one name="user" class="User" column="studentId" />
    </class>
</hibernate-mapping>
```

8.6.2 学生选题模块的实现

1．界面设计

学生选题界面的设计及操作如图 8.18 所示。

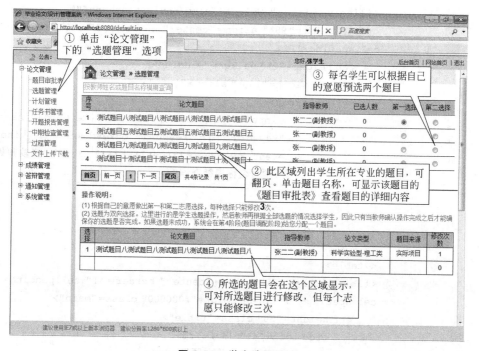

图 8.18 学生选题界面

2. 代码设计

```jsp
<%@page language="java" pageEncoding="utf-8" isELIgnored="false"%>
<!DOCTYPE HTML PUBLIC "-//W3C//DTD HTML 4.01 Transitional//EN">
<html>
    <head>
        <title>毕业论文(设计)管理系统</title>
        <%@include file="/WEB-INF/comm/public.jspf"%>
        <script type="text/javascript" src="${pageContext.request.contextPath}/js/list_select.js" charset="utf-8">
        </script>
        <style type="text/css">

        </style>
    </head>
    <body>
        <s:set var="choice1" value="#request.preelectionlist.get(0)"/>
        <s:set var="choice2" value="#request.preelectionlist.get(1)"/>
        <s:if test="#request.result!=null">
            <form action="preelectionAction_list.action" id="preelectionform" method="post">
                <s:hidden name="page" id="page"/>
                <%@include file="/WEB-INF/comm/navigation.jspf"%>
                <table width="98%" align="center" border="0" cellpadding="0" cellspacing="0">
                    <tr height="25" valign="middle">
                        <td width="100%" align="left">
                            <span>
                                <input type="text" value="${keyword}" name="keyword" id="keyword" onkeydown="if(window.event.keyCode==13)getlist();" title="按教师姓名或题目名称模糊查询"/>
                            </span>

                        </td>
                    </tr>
                </table>
                <table width="98%" align="center" border="1" cellpadding="0" cellspacing="0" bordercolor="#000000" class="main">
                    <tr height="30" class="title">
                        <td width="4%">序号</td>
                        <td width="51%">论文题目</td>
                        <td width="15%">指导教师</td>
                        <td width="12%">已选人数</td>
```

```
            <td width="9%">第一选择</td>
            <td width="9%">第二选择</td>
        </tr>
<s:set scope="pageContext" name="pagearg" value="((#request.result.currentpage-1) * #request.result.maxresult)"/>
<s:iterator value="#request.result.resultlist" status="st">
        <tr id='${id}' height="30" valign="middle" onmouseover="this.style.background='#eeeeee';" onmouseout="this.style.background='#FFFFFF';">
            <td align="center">
                <s:property value="#st.index+1+#pagearg"/>
            </td>
            <td align="left">
                <a href="subjectAction_detail.action?id=${id}" id='subjectname_${id}'><s:property value="name"/></a>
            </td>
            <td align="center" id='tea1_${id}'>
                <s:property value="tea1.name"/><s:property value="'('+tea1.Teachtitle.name+')'"/>
            </td>
            <td align="center">
                <s:if test="#request.nummap==null||#request.nummap.get(id)==null">
                    0
                </s:if>
                <s:else>
                    <s:property value="#request.nummap.get(id)"/>
                </s:else>
            </td>
            <td align="center">
                <input type="radio" name="1" id="1_${id}" onclick="preelection('${id}','1');" ${id==choice1.subject.id?'checked':''}
                    <s:if test="#request.times[0]==3">disabled="disabled"</s:if>
                />
            </td>
            <td align="center">
                <input type="radio" name="2" id="2_${id}" onclick="preelection('${id}','2');" ${id==choice2.subject.id?'checked':''}
                    <s:if test="#request.times[1]==3">disabled
```

```
                            ="disabled"</s:if>
                        />
                        <input type="hidden" id="type_${id}" value=
                        "${type.name}"/>
                        <input type="hidden" id="source_${id}" value=
                        "${source.name}"/>
                    </td>
                </tr>
            </s:iterator>
        </table>
    </form>
    <s:include value="/WEB-INF/comm/pagination.jsp"/>
    <hr align="center" width="100%" color="blur"/>
    <table width="98%" align="center" border="0" cellpadding="0"
    cellspacing="0" style="color: red;">
        <tr align="left">
            <td><b>操作说明:</b></td>
        </tr>
        <tr align="left">
            <td>(1)根据自己的意愿做出第一和第二志愿选择,每种选择只能修改
            <strong style="font-size:20px;color:#000000;"><!--#0d1ce9
            -->3</strong>次。</td>
        </tr>
        <tr align="left">
            <td>(2)选题为双向选择。这里进行的是学生选题操作,然后教师再根据
            全部选题的情况选择学生,因此只有当教师确认操作完成之后才能确保你
            的选题是否完成。如果选题未成功,系统会在第4阶段(题目调配阶段)给
            你分配一个题目。</td>
        </tr>
    </table>
</s:if>
<s:else>
    <table width="98%" align="center" border="0" cellpadding="0"
    cellspacing="0" style="color: red;">
        <tr align="left">
            <td><s:property value="#request.message"/></td><!--您的选
            题修改次数已达到上限,无法再次修改,请等待老师确认后的结果。-->
        </tr>
    </table>
</s:else>
<table width="98%" align="center" border="1" cellpadding="0"
cellspacing="0" bordercolor="#000000">
    <tr height="30" align="center" class="title">
        <td width="4%">选择</td>
```

```html
            <td width="47%">论文题目</td>
            <td width="14%">指导教师</td>
            <td width="18%">论文类型</td>
            <td width="10%">题目来源</td>
            <td width="7%">修改次数</td>
        </tr>
        <tr align="center" id='${id}' height="30" valign="middle" onmouseover="this.style.background='#eeeeee';" onmouseout="this.style.background='#FFFFFF';">
            <td id="choice_1"><s:property value="#choice1.choice"/></td>
            <td id="subjectname_1" align="left"><s:property value="#choice1.subject.name"/></td>
            <td id="tea1_1"><s:property value="#choice1.subject.tea1.name"/><s:property value="'('+#choice1.subject.tea1.Teachtitle.name+')'"/></td>
            <td id="type_1"><s:property value="#choice1.subject.type.name"/></td>
            <td id="source_1"><s:property value="#choice1.subject.source.name"/></td>
            <td id="times_1"><s:property value="#request.times[0]"/></td>
        </tr>
        <tr align="center" id='${id}' height="30" valign="middle" onmouseover="this.style.background='#eeeeee';" onmouseout="this.style.background='#FFFFFF';">
            <td id="choice_2"><s:property value="#choice2.choice"/></td>
            <td id="subjectname_2" align="left"><s:property value="#choice2.subject.name"/></td>
            <td id="tea1_2"><s:property value="#choice2.subject.tea1.name"/><s:property value="'('+#choice2.subject.tea1.Teachtitle.name+')'"/></td>
            <td id="type_2"><s:property value="#choice2.subject.type.name"/></td>
            <td id="source_2"><s:property value="#choice2.subject.source.name"/></td>
            <td id="times_2"><s:property value="#request.times[1]"/></td>
        </tr>
    </table>
    <input type="hidden" id="choosen_1" value="<s:property value='#choice1.subject.id'/>"/>
    <input type="hidden" id="choosen_2" value="<s:property value='#choice2.subject.id'/>"/>
    <div id="fbwindow" style="display:none;"></div>
</body>
</html>
```

8.6.3 教师确认学生子模块的实现

1. 界面设计

教师确认学生界面的设计及操作如图 8.19 所示。

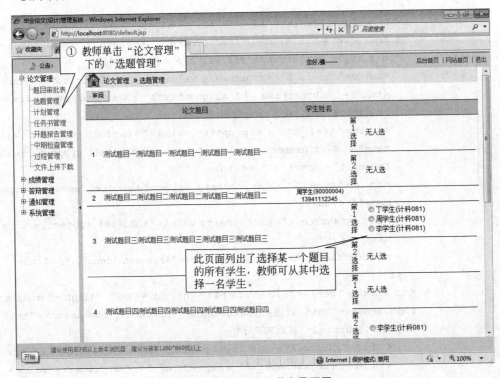

图 8.19 教师确认学生界面图

2. 代码设计

```
<%@page language="java" pageEncoding="utf-8" isELIgnored="false"%>
<!DOCTYPE HTML PUBLIC "-//W3C//DTD HTML 4.01 Transitional//EN">
<html>
    <head>
        <title>毕业论文(设计)管理系统</title>
        <%@include file="/WEB-INF/comm/public.jspf"%>
        <script type=" text/javascript" src=" ${pageContext. request.
contextPath}/js/list_confirm.js" charset="utf-8"></script>
        <style type="text/css">
            .page{
                border-bottom: 1px white solid;
            }
            .student{width:130px;display: inline;}
        </style>
```

```
</head>
<body>
    <form action =" subjectAction_confirmlist.action" id =" subjectform"
method="post">
        <%@include file="/WEB-INF/comm/privilege.jspf"%>
        <s:hidden name="page" id="page"/>
        <%@include file="/WEB-INF/comm/navigation.jspf"%>
        <s:if test="#session.user.role==0"><!--教师-->
            <table width="98%" align="center" border="0" cellpadding="0"
            cellspacing="0">
                <tr height="25" valign="middle">
                    <td width="100%" align="left">
                        <s:if test="#MAJOR||#COLLEGE||#SCHOOL">
                            <input type="button" value="审阅" class="btn2"
                            onclick="window.location.href='preelectionAction_
                            adminList.action?jsessionid = ${pageContext.
                            session.id}';"/>
                        </s:if>
                    </td>
                </tr>
            </table>
        </s:if>
        <table width =" 98%" align =" center" border =" 1" cellpadding =" 0"
        cellspacing="0" bordercolor="#000000" class="main">
            <tr height="30" class="title">
                <td width="4.5%">   </td>
                <td width="48.5%">论文题目</td>
                <td width="14%">学生姓名</td>
                <td width="33%">   </td>
            </tr>
            <s:set scope="pageContext" name="pagearg" value="((#request.
            result.currentpage-1) * #request.result.maxresult)"/>
            <s:bean name="edu.dldx.bylw.domain.Status" id="stat"/>
            <s:iterator value="#request.result.resultlist" status="st">
                <tr height="30" valign="middle" onmouseover="this.style.
                background='#eeeeee';" onmouseout="this.style.background=
                '#FFFFFF';">
                    <td align="center">
                        <s:property value="#st.index+1+#pagearg"/>
                    </td>
                    <td align="left">
                        <a href="subjectAction_detail.action?id=${id}">
                        ${name}</a><s:if test="#session.user.role==@edu.
                        dldx.bylw.domain.User@TEACHER&&#session.user.id.
```

```html
                    equals(tea2.id)">【第二指导】</s:if>
                </td>
                <td align="center" style="font-size:12px;" id="${id}" title="${student.clazz.name}">
                    ${student.name}<s:property value="'('+student.id+')'"/>
                    <br/>
                    ${student.tel}
                </td>
                <td align="center">
                    <s:if test="student==null">
                        <table width="100%" id="studentlist_${id}" border="0" style="font-size:12px; margin: 2px 0px;" bordercolor="#000000" cellpadding="0" cellspacing="0">
                            <tr height="40" valign="middle" align="left">
                                <td width="12%">第1<br/>选择</td>
                                <td width="88%">
                                    <s:if test="#request.map1.get(top.id)==null">无人选</s:if>
                                    <s:else>
                                        <s:iterator value="#request.map1.get(top.id)" var="first">
                                            <div class="student">
                                                <input type="radio" name="1" id='<s:property value="#first[0]"/>' onclick="confirmit(this,'${id}');"/><s:property value="#first[1]+'('+#first[2]+')'"/>
                                            </div>
                                        </s:iterator>
                                    </s:else>
                                </td>
                            </tr>
                            <tr height="1" valign="middle" align="left">
                                <td colspan="2">
                                    <hr width="100%" color="#000000" size="1" noshade="noshade"/>
                                </td>
                            </tr>
                            <tr height="40" valign="middle" align="left">
```

```
                        <td>第 2<br/>选择</td>
                        <td>
                            <s:if test="#request.map2.get(top.id)==null">无人选</s:if>
                            <s:else>
                                <s:iterator value="#request.map2.get(id)" var="second">
                                    <div class="student"><input type="radio" name="2" id='<s:property value="#second[0]"/>' onclick="confirmit(this,'${id}');"/><s:property value="#second[1]+'('+#second[2]+')'"/></div>
                                </s:iterator>
                            </s:else>
                        </td>
                    </tr>
                </table>
            </s:if>
            <s:else>${student.clazz.major.name} &raquo; ${student.clazz.name}</s:else>
        </td>
    </tr>
</s:iterator>
</table>
</form>
<s:include value="/WEB-INF/comm/pagination.jsp"/>
<div id="fbwindow" style="display:none;"></div>
</body>
</html>
```

8.6.4 题目调配子模块的实现

1. 界面设计

题目调配界面的设计及操作如图 8.20 所示。

2. 代码设计

```
<%@page language="java" pageEncoding="utf-8" isELIgnored="false"%>
<!DOCTYPE HTML PUBLIC "-//W3C//DTD HTML 4.01 Transitional//EN">
<html>
```

```
<head>
    <title>毕业论文(设计)管理系统</title>
    <%@ include file="/WEB-INF/comm/public.jspf"%>
    <script type=" text/javascript " src =" ${pageContext.request.contextPath}/js/adminList_preelection.js" charset="utf-8"></script>
    <style type="text/css">
        br{display:none;}
    </style>
    <script type="text/javascript">
        $(function(){
            $('#keyword').hide();
        });
    </script>
</head>
<body>
    <s:form action=" preelectionAction_adminList.action" id="preelectionform" method="post">
        <%@ include file="/WEB-INF/comm/privilege.jspf"%>
        <%@ include file="/WEB-INF/comm/navigation.jspf"%>
        <input type="hidden" name="which" id="which" value="${which}"/>
        <table width="98%" align="center" border="0" cellpadding="0" cellspacing="0">
            <tr height="25" valign="middle">
                <td width="100%" align="left">
                    <%@ include file="/WEB-INF/comm/admintitle.jspf"%>
                    <span>
                        <input type="button" class="btn2 L7" value="调配${which==0?'学生选题':'指导教师'}"
                            onclick="window.location.href='preelectionAction_adminList.action?jsessionid=${pageContext.session.id}&which=${which==0?1:0}';"
                            title="${which==1?'学生报的题目,未分配指导教师':'教师报的题目,没有学生选'}"
                        >
                    </span>
                    <span>
                        <input type="button" class="btn2 L7" value="下载选题统计表" onclick="download('subject','0001');">
                    </span>
                    <span id="0001"></span>
                </td>
            </tr>
```

```html
            </table>
            <table width =" 98%" align =" center" border =" 1" cellpadding =" 0"
cellspacing="0" bordercolor="#000000" class="main">
                <s:if test="#request.which==1">
                    <tr height="30" class="title">
                        <td width="30%" bgcolor="#dddddd">未选题学生()</td>
                        <td width =" 3 " style =" background - image: url
                        (${pageContext. request. contextPath }/images/blackline.
                        png);background-repeat:repeat-y;">
                        </td>
                        <td width="70%" bgcolor="#cccccc">剩余论文题目(个)</td>
                    </tr>
                    <tr height="30">
                        <td align="left" valign="top">
                            <s:iterator value="#request.studentlist" status=
                            "st">
                                <div style =" height: 25px; line - height: 25px;"
                                title="${tel}">
                                    <span style="width=20px;"><input type=
                                    "radio" name="stu" id="stu_${id}" value=
                                    "${id}" onclick="adjust();"/></span>
                                    <span style="width=70px;">${id}</span>
                                    <span style="width=45px;">${name}</span>
                                    <span style="width=60px;">${clazz.name}
</span>
                                </div>
                            </s:iterator>
                        </td>
                        <td width =" 3 " style =" background - image: url
                        (${pageContext. request. contextPath }/images/blackline.
                        png);background-repeat:repeat-y;">
                        </td>
                        <td align="left" valign="top">
                            <s:iterator value="#request.subjectlist" status=
                            "st">
                                <div style="height:25px;line-height:25px;">
                                    <span style="width=20px;"><input type=
"radio" name="sub" id="sub_${id}" value=
"${id}" onclick="adjust();"/></span>
                                    <span style="width=350px;"><a href=
"subjectAction_detail.action?id=${id}"> ${name}</a></span>
                                    <span style =" width = 45px;" > ${ teal. name }
```

```
                            </span>
                            <span style="width=120px;">${tea1.office.
                            name}</span>
                        </div>
                    </s:iterator>
                </td>
            </tr>
        </s:if>
        <s:elseif test="#request.which==0">
            <tr height="30" class="title">
                <td width="3%">  </td>
                <td width="45%">论文题目</td>
                <td width="12%">指导教师</td>
                <td width="14%">学生姓名</td>
                <td width="18%">状态</td>
                <td width="8%">   </td>
            </tr>
            <s:bean name="edu.dldx.bylw.domain.Status" id="stat"/>
            <s:iterator value="#request.subjectlist" status="st" var=
            "subject">
                <tr height="30" valign="middle" onmouseover="this.
                style.background='#eeeeee';" onmouseout="this.style.
                background='#FFFFFF';">
                    <td>
                        <img src="${pageContext.request.contextPath}/
                        images/left_notice.gif"/>
                    </td>
                    <td align="left"><a href="subjectAction_detail.
                    action?id=${id}">${name}</a></td>
                    <td align="center">
                        <s:select list="#request.teacherlist" headerKey
                        ="-1" headerValue="请选择" listKey="id"
                        listValue="name+'('+#request.map.get(id)+')'
                        "name="tea1Id" id="tea1Id_%{id}" onchange=
                        "adjust_tea('%{id}');"/>
                    </td>
                    <td align="center" style="font-size:12px;" title=
                    "${student.clazz.name}">
                        <p>${student.name}<s:property value="'('+
                        student.id+')'"/></p>
                        ${student.tel}
                    </td>
```

```xml
                    <td align="center">
                        <s:property value="#stat.getDetail(status)"/>
                    </td>
                    <td id="${id}">   </td>
                </tr>
            </s:iterator>
        </s:elseif>
    </table>
</s:form>
<div id="fbwindow" style="display:none;"></div>
</body>
</html>
```

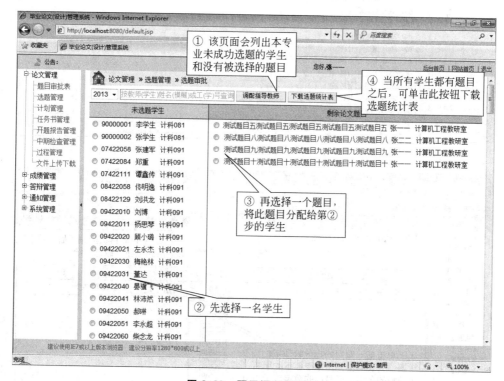

图 8.20　题目调配界面图

8.6.5　调配教师子模块的实现

1. 界面设计

调配教师界面的设计及操作如图 8.21 所示。

2. 代码设计

参考 8.7.4 节的代码。

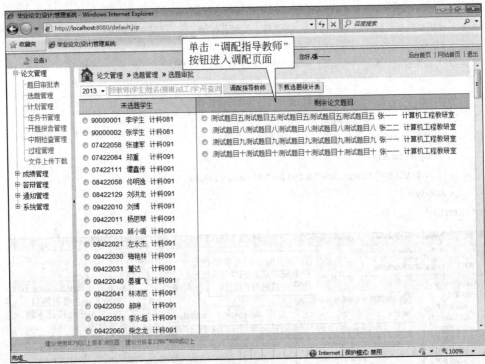

(a) 调配教师界面1

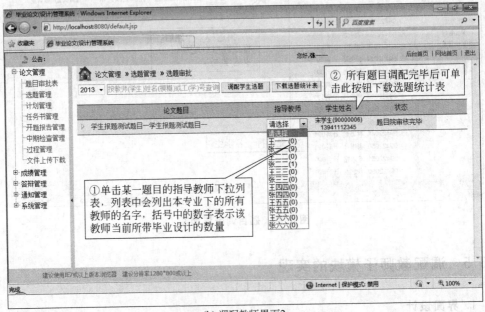

(b) 调配教师界面2

图 8.21 调配教师界面的设计

参 考 文 献

[1] 董眯芳,刘振安. UML 课程设计[M]. 北京：机械工业出版社,2006.
[2] 邱少明,袁劲松. Java 语言程序设计[M]. 北京：清华大学出版社,北京交通大学出版社,2009.
[3] 韩万江,江立新,等. 软件工程案例教程[M]. 2 版. 北京：机械工业出版社,2013.
[4] http://baike.baidu.com/view/589832.htm.
[5] http://baike.baidu.com/view/1317054.htm.
[6] http://baike.baidu.com/view/1488767.htm.
[7] http://baike.baidu.com/view/2814288.htm.
[8] http://baike.baidu.com/view/1938914.htm.
[9] http://baike.baidu.com/view/977673.htm.
[10] (英)Ian Sommerville 著. 软件工程(原书第 9 版)[M]. 北京：机械工业出版社,2011.
[11] 张海潘. 软件工程导轮(第五版)[M]. 北京：清华大学出版社,2010.
[12] 麻志毅. 面向对象开发方法[M]. 北京：机械工业出版社,2011.

参考文献

[1] 柳冠中. 事理学论纲[M]. 长沙: 中南大学出版社, 2006.
[2] 李乐山. 工业设计思想基础[M]. 北京: 中国建筑工业出版社, 2007.
[3] 李砚祖. 艺术设计概论[M]. 武汉: 湖北美术出版社, 2012.
[4] http: //baike.baidu.com/view/763680.htm.
[5] http: //baike.baidu.com/view/517058.htm.
[6] http: //baike.baidu.com/view/28709.htm.
[7] http: //baike.baidu.com/view/2818285.htm.
[8] http: //baike.baidu.com/view/95561.htm.
[9] http: //baike.baidu.com/view/1073919.htm.
[10] 克里斯Sommer. 设计的立体化[M]. 武汉: 华中科技大学出版社, 2011.
[11] 鲁道夫.阿恩海姆. 艺术与视知觉[M]. 成都: 四川人民出版社, 2010.
[12] 鲁晓波. 信息产品设计[M]. 北京: 清华大学出版社, 2004.